ENTANGLEMENT

The Greatest Mystery in Physics

AMIR D. ACZEL

FOUR WALLS EIGHT WINDOWS
NEW YORK

Published in the United States by:
Four Walls Eight Windows
39 West 14th Street, room 503
New York, N.Y., 10011

Visit our website at http://www.4w8w.com

First printing September 2002.

Library of Congress Cataloging-in-Publication Data:

    Entanglement: the greatest mystery in physics/ by Amir D. Aczel.
      p.   cm.
    Includes bibliographical references and index.
      ISBN 1-56858-232-3
      1. Quantum theory. I. Title.
    QC174.12.A29 2002
    530.12—dc21   2002069338

10 9 8 7 6 5 4 3 2 1

Printed in the United States

Typeset and designed by Terry Bain

Illustrations, unless otherwise noted, by Ortelius Design.

for Ilana

Contents

Preface

"My own suspicion is that the universe is not only queerer than we suppose, but queerer than we can suppose."
—J.B.S. Haldane

*I*n the fall of 1972, I was an undergraduate in mathematics and physics at the University of California at Berkeley. There I had the good fortune to attend a special lecture given on campus by Werner Heisenberg, one of the founders of the quantum theory. While today I have some reservations about the role Heisenberg played in history—at the time other scientists left in protest of Nazi policies, he stayed behind and was instrumental in Hitler's attempts to develop the Bomb—nevertheless his talk had a profound, positive effect on my life, for it gave me a deep appreciation for the quantum theory and its place in our efforts to understand nature.

Quantum mechanics is the strangest field in all of science. From our everyday perspective of life on Earth, nothing makes sense in quantum theory, the theory about the laws of nature that govern the realm of the very small (as well as some large systems, such as superconductors). The word itself, *quantum*, denotes a small packet of energy—a very small one. In quantum mechanics, as the quantum theory is called, we deal with the basic building blocks of matter, the constituent particles from which everything in the universe is made. These particles include atoms, molecules, neutrons, protons, electrons, quarks, as well as photons—the basic units of light. All these objects (if indeed they can be called objects) are much smaller than anything the human eye can see. At this level, suddenly, all the rules of behavior with which we are familiar no longer hold. Entering this strange new world of the very small is an experience as baffling and bizarre as Alice's adventures in Wonderland. In this unreal quantum world, particles are waves, and waves are particles. A ray of light, therefore, is both an electromagnetic wave undulating through space, and a stream of tiny particles speeding toward the observer, in the sense that some quantum experiments or phenomena reveal the wave nature of light, while others the particle nature of the same light—but never both aspects at the same time. And yet, before we observe a ray of light, it is both a wave and a stream of particles.

In the quantum realm everything is fuzzy—there is a hazy quality to all the entities we deal with, be they light or electrons or atoms or quarks. An *uncertainty principle* reigns in quantum mechanics, where most things cannot be seen or felt or known with precision, but only through a haze of probability and chance. Scientific predictions about outcomes are statistical in nature and are given in terms of probabilities—we can only predict the most likely location of a particle, not its exact position. And we can never determine both a particle's location and its momentum with good accuracy. Furthermore, this fog that permeates the quantum world can never go away. There are no "hidden variables," which, if known, would increase our precision beyond the natural limit that rules the quantum world. The uncertainty, the fuzziness, the probabilities, the dispersion simply cannot go away—these mysterious, ambiguous, veiled elements are an integral part of this wonderland.

Even more inexplicable is the mysterious *superposition* of states of quantum systems. An electron (a negatively-charged elementary particle) or photon (a quantum of light) can be in a superposition of two or more states. No longer do we speak about "here or there;" in the quantum world we speak about "here *and* there." In a certain sense, a photon, part of a stream of light shone on a screen with two holes, can go through *both* holes at the same time, rather than the expected choice of one hole *or* the other. The electron in orbit around the nucleus is potentially at many locations at the same time.

But the most perplexing phenomenon in the bizarre world of the quantum is the effect called *entanglement*. Two particles that may be very far apart, even millions or billions of miles, are mysteriously linked together. Whatever happens to one of them *immediately* causes a change in the other one.[1]

What I learned from Heisenberg's lecture thirty years ago was that we must let go of all our preconceptions about the world derived from our experience and our senses, and instead let mathematics lead the way. The electron lives in a different space from the one in which we live. It lives in what mathematicians call a *Hilbert space*, and so do the other tiny particles and photons. This Hilbert space, developed by mathematicians independently of physics, seems to describe well the mysterious rules of the quantum world—rules that make no sense when viewed with an eye trained by our everyday experiences. So the physicist working with quantum systems relies on the mathematics to produce predictions of the outcomes of experiments or phenomena, since this same physicist has no natural intuition about what goes on inside an atom or a ray of light or a stream of particles. Quantum theory taxes our very concept of what constitutes science— for we can never truly "understand" the bizarre behavior of the very small. And it taxes our very idea of what constitutes reality. What does "reality" mean in the context of the existence of entangled entities that act in concert even while vast distances apart?

The beautiful mathematical theory of Hilbert space, abstract algebra, and probability theory—our mathematical tools for handling quantum phenomena—allow us to *predict* the results of experiments to a stunning level of accuracy; but they do not bring us an *understanding* of the underlying processes. Understanding what really happens inside the mysterious box constituting a quantum system may be beyond the powers of human beings. According to one interpretation of quantum mechanics, we can only use the box to predict outcomes. And these predictions are statistical in nature.

There is a very strong temptation to say: "Well, if the

theory cannot help us understand what truly goes on, then the theory is simply not *complete*. Something is missing—there must be some missing variables, which, once added to our equations, would complete our knowledge and bring us the understanding we seek." And, in fact, the greatest scientist of the twentieth century, Albert Einstein, posed this very challenge to the nascent quantum theory. Einstein, whose theories of relativity revolutionized the way we view space and time, argued that quantum mechanics was excellent as a statistical theory, but did not constitute a complete description of physical reality. His well-known statement that "God doesn't play dice with the world" was a reflection of his belief that there was a deeper, non-probabilistic layer to the quantum theory which had yet to be discovered. Together with his colleagues Podolsky and Rosen, he issued a challenge to quantum physics in 1935, claiming that the theory, was incomplete. The three scientists based their argument on the existence of the entanglement phenomenon, which in turn had been deduced to exist based on mathematical considerations of quantum systems.

At his talk at Berkeley in 1972, Heisenberg told the story of his development of the approach to the quantum theory called *matrix mechanics*. This was one of his two major contributions to the quantum theory, the other being the uncertainty principle. Heisenberg recounted how, when aiming to develop his matrix approach in 1925, he did not even know how to multiply matrices (an elementary operation in mathematics). But he taught himself how to do so, and his theory followed. Mathematics thus gave scientists the rules of behavior in the quantum world. Mathematics also led Schrödinger to his alternative, and simpler, approach to quantum mechanics, the wave equation.

Over the years, I've followed closely the developments in the quantum theory. My books have dealt with mysteries in mathematics and physics. *Fermat's Last Theorem* told the story of the amazing proof of a problem posed long ago; *God's Equation* was the tale of Einstein's cosmological constant and the expansion of the universe; *The Mystery of the Aleph* was a description of humanity's attempt to understand infinity. But I've always wanted to address the secrets of the quantum. A recent article in *The New York Times* provided me with the impetus I needed. The article dealt with the challenge Albert Einstein and his two colleagues issued to the quantum theory, claiming that a theory that allowed for the "unreal" phenomenon of entanglement had to be incomplete.

Seven decades ago, Einstein and his scientific allies imagined ways to prove that quantum mechanics, the strange rules that describe the world of the very small, were just too spooky to be true. Among other things, Einstein showed that, according to quantum mechanics, measuring one particle could instantly change the properties of another particle, no matter how far apart they were. He considered this apparent action-at-a-distance, called entanglement, too absurd to be found in nature, and he wielded his thought experiments like a weapon to expose the strange implications that this process would have if it could happen. But experiments described in three forthcoming papers in the journal Physical Review Letters give a measure of just how badly Einstein has been routed. The

experiments show not only that entanglement does happen—
which has been known for some time—but that it might be
used to create unbreakable codes . . .[2]

As I knew from my study of the life and work of Albert Einstein, even when Einstein thought he was wrong (about the
cosmological constant), he was right. And as for the quantum
world—Einstein was one of the developers of the theory. I
knew quite well that—far from being wrong—Einstein's
paper of 1935, obliquely alluded to in the *Times* article, was
actually the seed for one of the most important discoveries in
physics in the twentieth century: the actual discovery of
entanglement through physical experiments. This book tells
the story of the human quest for entanglement, the most
bizarre of all the strange aspects of quantum theory.

Entangled entities (particles or photons) are linked together
because they were produced by some process that bound
them together in a special way. For example, two photons
emitted from the same atom as one of its electrons descends
down two energy levels are entangled. (Energy levels are
associated with the orbit of an electron in the atom.) While
neither flies off in a definite direction, the pair will always be
found on opposite sides of the atom. And such photons or
particles, produced in a way that links them together, remain
intertwined forever. Once one is changed, its twin—*wherever it may be in the universe*—will change *instantaneously.*

In 1935, Einstein, together with his colleagues Rosen and
Podolsky, considered a system of two distinct particles that
was permissible under the rules of quantum mechanics. The
state of this system was shown to be entangled. Einstein,

Podolsky, and Rosen used this theoretical entanglement of separated particles to imply that if quantum mechanics allowed such bizarre effects to exist, then something must be wrong, or *incomplete*, as they put it, about the theory.

In 1957, the physicists David Bohm and Yakir Aharonov analyzed the results of an experiment that had been performed by C.S. Wu and I. Shaknov almost a decade earlier, and their analysis provided the first hint that entanglement of separated systems may indeed take place in nature. Then in 1972, two American physicists, John Clauser and Stuart Freedman, produced evidence that entanglement actually exists. And a few years later, the French physicist Alain Aspect and his colleagues provided more convincing and complete evidence for the existence of the phenomenon. Both groups followed the seminal theoretical work in this area by John S. Bell, an Irish physicist working in Geneva, and set out to prove that the Einstein-Podolsky-Rosen thought experiment was not an absurd idea to be used to invalidate the completeness of the quantum theory, but rather the description of a real phenomenon. The existence of the phenomenon provides evidence in favor of quantum mechanics and against a limiting view of reality.

A NOTE TO THE READER

Quantum theory itself, and in particular the concept of entanglement, is very difficult for anyone to understand—even for accomplished physicists or mathematicians. I therefore structured the book in such a way that the ideas and concepts discussed are constantly being explained and re-explained in various forms. This approach makes sense when one considers the fact that some of the brightest scientists today have spent lifetimes working on entanglement; the truth is that even after decades of research, it is difficult to find someone who will admit to understanding the quantum theory perfectly well. These physicists know how to apply the rules of quantum mechanics in a variety of situations. They can perform calculations and make predictions to a very high degree of accuracy, which is rare in some other areas. But often these bright scientists will profess that they do not truly *understand* what goes on in the quantum world. It is exactly for this reason that in chapter after chapter in this book I repeat the concepts of quantum theory and entanglement, every time from a slightly different angle, or as explained by a different scientist.

I have made an effort to incorporate the largest possible number of original figures, obtained from scientists, describing actual experiments and designs. My hope is that these figures and graphs will help the reader understand the mysterious and wonderful world of the quantum and the setting within which entanglement is produced and studied. In addition, where appropriate, I have incorporated a number of equations and symbols. I did so not to baffle the reader, but so that readers with an advanced preparation in science might gain more from the presentation. For example, in the chapter on Schrödinger's work I include the simplest (and most restricted) form of Schrödinger's famous equation for the benefit of those who might want to see what the equation looks like. It is perfectly fine for a reader, if she so chooses, to skip over the equations and read on, and anyone doing so will suffer no loss of information or continuity.

This is a book about *science*, the making of science, the philosophy that underlies science, the mathematical underpinnings of science, the experiments that verify and expose nature's inner secrets, and the lives of the scientists who pursue nature's most bizarre effect. These scientists constitute a group of the greatest minds of the twentieth century, and their combined lifetimes span the entire century. These people, relentlessly in search of knowledge about a deep mystery of nature—entanglement—led and lead lives today that are, themselves, entangled with one another. This book tells the story of this search, one of the greatest scientific detective stories in history. And while the science of entanglement has also brought about the birth of new and very exciting technologies, the focus of this book is not on the technologies spawned by the research. *Entanglement* is about the search called modern science.

1

A Mysterious Force of Harmony

"Alas, to wear the mantle of Galileo it is not enough that you be persecuted by an unkind establishment, you must also be right."

—Robert Park

*I*s it possible that something that happens here will *instantaneously* make something happen at a far away location? If we measure something in a lab, is it possible that at the same moment, a similar event takes place ten miles away, on the other side of the world, or on the other side of the universe? Surprisingly, and against every intuition we may possess about the workings of the universe, the answer is *yes*. This book tells the story of *entanglement*, a phenomenon in which two entities are inexorably linked no matter how far away from each other they may be. It is the story of the people who have spent lifetimes seeking evidence that such a bizarre effect—predicted by the quantum theory and brought to wide scientific attention by Einstein—is indeed an integral part of nature.

As these scientists studied such effects, and produced definitive evidence that entanglement is a reality, they have also discovered other, equally perplexing, aspects of the phenomenon. Imagine Alice and Bob, two happily married people. While Alice is away on a business trip, Bob meets Carol, who is married to Dave. Dave is also away at that time, on the other side of the world and nowhere near any of the other three. Bob and Carol become entangled with each other; they forget their respective spouses and now strongly feel that they are meant to stay a couple forever. Mysteriously, Alice and Dave—who have never met—are now also entangled with each other. They suddenly share things that married people do, without ever having met. If you substitute for the people in this story particles labeled A, B, C, and D, then the bizarre outcome above actually occurs. If particles A and B are entangled, and so are C with D, then we can entangle the separated particles A and D by passing B and C through an apparatus that entangles them together.

Using entanglement, the state of a particle can also be *teleported* to a faraway destination, as happens to Captain Kirk on the television series "Star Trek" when he asks to be beamed back up to the *Enterprise*. To be sure, no one has yet been able to teleport a person. But the state of a quantum system has been teleported in the laboratory. Furthermore, such incredible phenomena can now be used in cryptography and computing.

In such futuristic applications of technology, the entanglement is often extended to more than two particles. It is possible to create triples of particles, for example, such that all three are 100% correlated with each other—whatever happens to one particle causes a similar instantaneous change in the other two. The three entities are thus inexorably inter-linked, wherever they may be.

One day in 1968, physicist Abner Shimony was sitting in his office at Boston University. His attention was pulled, as if by a mysterious force, to a paper that had appeared two years earlier in a little-known physics journal. Its author was John Bell, an Irish physicist working in Geneva. Shimony was one of very few people who had both the ability and the desire to truly understand Bell's ideas. He knew that Bell's theorem, as explained and proved in the paper, allowed for the possibility of testing whether two particles, located far apart from each other, could act in concert. Shimony had just been asked by a fellow professor at Boston University, Charles Willis, if he would be willing to direct a new doctoral student, Michael Horne, in a thesis on statistical mechanics. Shimony agreed to see the student, but was not eager to take on a Ph.D. student in his first year of teaching at Boston University. In any case, he said, he had no good problem to suggest in statistical mechanics. But, thinking that Horne might find a problem in the foundations of quantum mechanics interesting, he handed him Bell's paper. As Shimony put it, "Horne was bright enough to see quickly that Bell's problem was interesting." Michael Horne took Bell's paper home to study, and began work on the design of an experiment that would use Bell's theorem.

Unbeknownst to the two physicists in Boston, at Columbia University in New York, John F. Clauser was reading the same paper by Bell. He, too, was mysteriously drawn to the problem suggested by Bell, and recognized the opportunity for an actual experiment. Clauser had read the paper by Einstein, Podolsky, and Rosen, and thought that their suggestion was very plausible. Bell's theorem showed a discrepancy between quantum mechanics and the "local hidden variables" interpretation of quantum mechanics offered by Einstein and his colleagues as an alternative to the "incomplete" quantum theory, and Clauser was excited about the possibility of an experiment exploiting this discrepancy. Clauser was skeptical, but he couldn't resist testing Bell's predictions. He was a graduate student, and everyone he talked to told him to leave it alone, to get his Ph.D., and not to dabble in science fiction. But Clauser knew better. The key to quantum mechanics was hidden within Bell's paper, and Clauser was determined to find it.

Across the Atlantic, a few years later, Alain Aspect was feverishly working in his lab in the basement of the Center for Research on Optics of the University of Paris in Orsay. He was racing to construct an ingenious experiment: one that would prove that two photons, at two opposite sides of his lab, could instantaneously affect each other. Aspect was led to his ideas by the same abstruse paper by John Bell.

In Geneva, Nicholas Gisin met John Bell, read his papers and was also thinking about Bell's ideas. He, too, was in the race to find an answer to the same crucial question: a question that had deep implications about the very nature of reality. But we are getting ahead of ourselves. The story of Bell's ideas, which goes back to a suggestion made thirty-five years earlier by Albert Einstein, has its origins in humanity's quest for knowledge of the physical world. And in order to truly understand these deep ideas, we must return to the past.

2

Before the Beginning

"Out yonder there was this huge world, which exists independently of us human beings and which stands before us like a great, eternal riddle, at least partially accessible to our inspection."

—Albert Einstein

"The mathematics of quantum mechanics is straightforward, but making the connection between the mathematics and an intuitive picture of the physical world is very hard"

—Claude N. Cohen-Tannoudji

*I*n the book of Genesis we read: "God said: Let there be light." God then created heaven and earth and all things that filled them. Humanity's quest for an understanding of light and matter goes back to the dawn of civilization; they are the most basic elements of the human experience. And, as Einstein showed us, the two are one and the same: both light and matter are forms of energy. People have always striven to understand what these forms of energy mean. What is the nature of matter? And what is light?

The ancient Egyptians and Babylonians and their successors the Phoenicians and the Greeks tried to understand the mysteries of matter, and of light and sight and color. The Greeks looked at the world with the first modern intellectual eyes. With their curiosity about numbers and geometry, coupled with a deep desire to understand the inner workings of nature and their environment, they gave the world its first ideas about physics and logic.

To Aristotle (300 B.C.), the sun was a perfect circle in the sky, with no blemishes or imperfections. Eratosthenes of Cyrene (c. 276 B.C.-194 B.C.) estimated the circumference of our planet by measuring the angle sunlight was making at Syene (modern Aswan), in Upper Egypt, against the angle it made at the same time farther north, in Alexandria. He came stunningly close to the earth's actual circumference of 25,000 miles.

The Greek philosophers Aristotle and Pythagoras wrote about light and its perceived properties; they were fascinated by the phenomenon. But the Phoenicians were the first people in history to make glass lenses, which allowed them to magnify objects and to focus light rays. Archaeologists have found 3,000-year-old magnifying glasses in the region of the eastern Mediterranean that was once Phoenicia. Interestingly, the principle that makes a lens work is the slowing-down of light as it travels through glass.

The Romans learned glass-making from the Phoenicians, and their own glassworks became one of the important industries of the ancient world. Roman glass was of high quality and was even used for making prisms. Seneca (5 B.C.-A.D. 45) was the first to describe a prism and the breaking-down of white light into its component colors. This phenomenon, too, is based on the speed of light. We have no evidence of any experiments carried out in antiquity to determine the speed of light. It seems that ancient peoples thought that light moved instantly from place to place. Because light is so fast, they could not detect the infinitesimal delays as light traveled from source to destination. The first attempt to study the speed of light did not come for another 1,600 years.

Galileo was the first person known to have attempted to estimate the speed of light. Once again, experimentation with light had a close connection with glassmaking. After the Roman Empire collapsed in the fifth century, many Romans of patrician and professional backgrounds escaped to the Venetian lagoons and established the republic of Venice. They brought with them the art of making glass, and thus the glassworks on the island of Murano were established. Galileo's telescopes were of such high quality—in fact they were far better than the first telescopes made in Holland— because he used lenses made of Murano glass. It was with the help of these telescopes that he discovered the moons of Jupiter and the rings of Saturn and determined that the Milky Way is a large collection of stars.

In 1607, Galileo conducted an experiment on two hilltops in Italy, in which a lantern on one hill was uncovered. When an assistant on the other hilltop saw the light, he opened his own lantern. The person on the first hill tried to estimate the time between opening the first lantern and seeing the light return from the second one. Galileo's quaint experiment failed, however, because of the tiny length of time elapsed between the sending of the first lantern signal and the return of the light from the other hilltop. It should be noted, anyway, that much of this time interval was due to the human response time in uncovering the second lantern rather than to the actual time light took to travel this distance.

Almost seventy years later, in 1676, the Danish astronomer Olaf Römer became the first scientist to calculate the speed of light. He accomplished this task by using astronomical observations of the moons of Jupiter, discovered by Galileo. Römer devised an intricate and extremely clever scheme by which he recorded the times of the eclipses of the moons of Jupiter. He knew that the earth orbits the sun, and that therefore the earth would be at different locations in space vis-a-vis Jupiter and its moons. Römer noticed that the times of disappearance of the moons of Jupiter behind the planet were not evenly spaced. As Earth and Jupiter orbit the sun, their distance from each other varies. Thus the light that brings us information on an eclipse of a Jovian moon takes different lengths of time to arrive on Earth. From these differences, and using his understanding of the orbits of Earth and Jupiter, Römer was able to calculate the speed of light. His estimate, 140,000 miles per second, was not quite the actual value of 186,000 miles per second. However, considering the date of the discovery and the fact that time was not measurable to great accuracy using the clocks of the seventeenth century, his achievement—the first measurement of the speed of light and the first proof that light does not travel at infinite speed—is an immensely valuable landmark in the history of science.

Descartes wrote about optics in 1638 in his book *Dioptrics*, stating laws of the propagation of light: the laws of reflection and refraction. His work contained the seed of the most controversial idea in the field of physics: the *ether*. Descartes put forward the hypothesis that light propagates through a medium, and he named this medium the ether. Science would not be rid of the ether for another three hundred years, until Einstein's theory of relativity would finally deal the ether its fatal blow.

Christiaan Huygens (1629-1695) and Robert Hooke (1635-1703) proposed the theory that light is a wave. Huygens, who as a sixteen-year-old boy had been tutored by Descartes during his stay in Holland, became one of the greatest thinkers of the day. He developed the first pendulum clock and did other work in mechanics. His most remarkable achievement, however, was a theory about the nature of light. Huygens interpreted Römer's discovery of the finite speed of light as implying that light must be a wave propagating through some medium. On this hypothesis, Huygens constructed an entire theory. Huygens visualized the medium as the ether, composed of an immense number of tiny, elastic particles. When these particles were excited into vibration, they produced light waves.

In 1692, Isaac Newton (1643-1727) finished his book *Opticks* about the nature and propagation of light. The book was lost in a fire in his house, so Newton rewrote it for publication in 1704. His book issued a scathing attack on Huygens's theory, and argued that light was not a wave but instead was composed of tiny particles traveling at speeds that depend on the color of the light. According to Newton, there are seven colors in the rainbow: red, yellow, green, blue, violet, orange, and indigo. Each color has its own speed of propagation. Newton derived his seven colors by an analogy with the seven main intervals of the musical octave. Further editions of his book continued Newton's attacks on Huygens's theories and intensified the debate as to whether light is a particle or a wave. Surprisingly, Newton—who co-discovered the calculus and was one of the greatest mathematicians of all time—never bothered to address Römer's findings about the speed of light, and neither did he give the wave theory the attention it deserved.

But Newton, building on the foundation laid by Descartes, Galileo, Kepler, and Copernicus, gave the world classical mechanics, and, through it, the concept of causality. Newton's second law says that force is equal to mass times acceleration: $F=ma$. Acceleration is the second derivative of position (it is the rate of change of the speed; and speed, in turn, is the rate of change of position). Newton's law is therefore an equation with a (second) derivative in it. It is called a (second-order) differential equation. Differential equations are very important in physics, since they model change. Newton's laws of motion are a statement about *causality*. They deal with cause and effect. If we know the initial position and velocity of a massive body, and we know the force acting on it and the force's direction, then we should be able to determine a final outcome: where will the body be at a later point in time.

Newton's beautiful theory of mechanics can predict the motion of falling bodies as well as the orbits of planets. We can use these cause-and-effect relationships to predict where an object will go. Newton's theory is a tremendous edifice that explains how large bodies—things we know from everyday life—can move from place to place, as long as their speeds or masses are not too great. For velocities approaching the speed of light, or masses of the order of magnitude of stars, Einstein's general relativity is the correct theory, and classical,

Newtonian mechanics breaks down. It should be noted, however, that Einstein's theories of special and general relativity hold, with improvements over Newton, even in situations in which Newtonian mechanics is a good approximation. Similarly, for objects that are very small—electrons, atoms, photons—Newton's theory breaks down as well. With it, we also lose the concept of causality. The quantum universe does not possess the cause-and-effect structure we know from everyday life. Incidentally, for small particles moving at speeds close to that of light, *relativistic* quantum mechanics is the right theory.

One of the most important principles in classical physics—and one that has great relevance to our story—is the principle of conservation of momentum. Conservation principles for physical quantities have been known to physicists for over three centuries. In his book, the *Principia*, of 1687, Newton presented his laws for the conservation of mass and momentum. In 1840, the German physician Julius Robert Mayer (1812-1878) deduced that energy was conserved as well. Mayer was working as a ship's surgeon on a voyage from Germany to Java. While treating members of the ship's crew for various injuries in the tropics, Dr. Mayer noticed that the blood oozing from their wounds was *redder* than the blood he saw in Germany. Mayer had heard of Lavoisier's theory that body heat came from the oxidation of sugar in body tissue using oxygen from the blood. He reasoned that in the warm tropics the human body needed to produce less heat than it would in colder northern Europe, and hence that more oxygen remained in the blood of people in the tropics, making the blood redder. Using arguments about how the body interacts with the environment—giving and receiving heat—Mayer postulated that energy was conserved. This idea was derived experimentally by Joule, Kelvin, and Carnot. Earlier, Leibniz had discovered that kinetic energy can be transformed into potential energy and vice versa.

Energy in any of its forms (including mass) is conserved—that is, it cannot be created out of nothing. The same holds true for momentum, angular momentum, and electric charge. The conservation of momentum is very important to our story.

Suppose that a moving billiards ball hits a stationary one. The moving ball has a particular *momentum* associated with it—the product of its mass by its speed, $p=mv$. This product of mass times speed, the momentum of the billiards ball, must be conserved within the system. Once one ball hits another, its speed slows down, but the ball that was hit now moves as well. The speed times mass for the system of these two objects must be the same as that of the system before the collision (the stationary ball had momentum zero, so it's the momentum of the moving one that now gets split in two). This is demonstrated by the figure below, where after the collision the two balls travel in different directions.

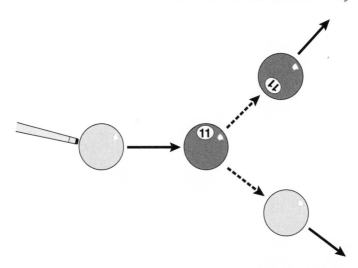

In any physical process, total input momentum equals total output momentum. This principle, when applied within the world of the very small, will have consequences beyond this simple and intuitive idea of conservation. In quantum mechanics, two particles that interact with each other at some point—in a sense like the two billiards balls of this example—will remain intertwined with each other, but to a greater extent than billiards balls: whatever should happen to one of them, no matter how far it may be from its twin, will immediately affect the twin particle.

3

Thomas Young's Experiment

"We choose to examine a phenomenon (the double-slit experiment) that is impossible, *absolutely* impossible, to explain in any classical way, and which has in it the heart of quantum mechanics. In reality, it contains the *only* mystery."

—Richard Feynman

Thomas Young (1773-1829) was a British physician and physicist whose experiment changed the way we think about light. Young was a child prodigy who learned to read at age two, and by age six had read the Bible twice and learned Latin. Before the age of 19, he was fluent in thirteen languages, including Greek, French, Italian, Hebrew, Chaldean, Syriac, Samaritan, Persian, Ethiopic, Arabic, and Turkish. He studied Newton's calculus and his works on mechanics and optics, as well as Lavoisier's *Elements of Chemistry*. He also read plays, studied law, and learned politics.

In the late 1700s Young studied medicine in London, Edinburgh, and Göttingen, where he received his M.D. In 1794, he was elected to the Royal Society. Three years later, he moved to Cambridge University, where he received a second

M.D. and joined the Royal College of Physicians. After a wealthy uncle left him a house in London and a large cash inheritance, Young moved to the capital and established a medical practice there. He was not a successful doctor, but instead devoted his energies to study and scientific experiments. Young studied vision and gave us the theory that the eye contains three types of receptors for light of the three basic colors, red, blue, and green. Young contributed to natural philosophy, physiological optics, and was one of the first to translate Egyptian hieroglyphics. His greatest contribution to physics was his effort to win acceptance of the wave theory of light. Young conducted the now-famous double-slit experiment on light, demonstrating the wave-theory effect of interference.

In his experiment, Young had a light source and a barrier. He cut two slits in the barrier, through which the light from the source could pass. Then he placed a screen behind the barrier. When Young shone the light from the source on the barrier with the two slits, he obtained an *interference* pattern.

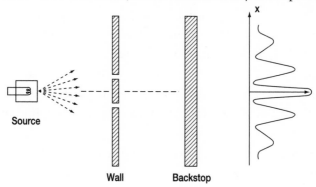

An interference pattern is the hallmark of waves. Waves interfere with each other, while particles do not. Richard Feynman considered Young's result of the double-slit experiment—as it appears in the case of electrons and other quanta that can be localized—so important that he devoted much of the first chapter of the third volume of his renowned textbook, *The Feynman Lectures on Physics*, to this type of experiment.[3] He believed that the result of the double-slit experiment was the fundamental mystery of quantum mechanics. Richard Feynman demonstrated in his *Lectures* the idea of interference of waves versus the non-interference of particles using bullets. Suppose a gun shoots bullets randomly at a barrier with two slits. The pattern is as shown below.

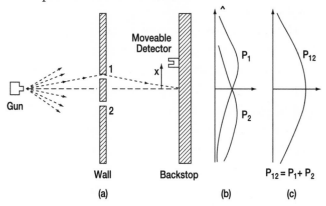

Water waves, if passed through a barrier with two slits, make the pattern below. Here we find interference, as in the Young experiment with light, because we have classical waves. The amplitudes of two waves may add to each other, producing a peak on the screen, or they may interfere destructively, producing a trough.

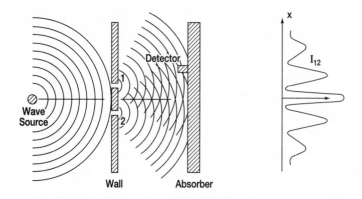

So the Young experiment demonstrates that light is a wave. But is light really a wave?

The duality between light as wave and light as a stream of particles still remains an important facet of physics in the twenty-first century. Quantum mechanics, developed in the 1920s and 1930s, in fact reinforces the view that light is both particle and wave. The French physicist Louis de Broglie argued in 1924 that even physical bodies such as electrons and other particles possess wave properties. Experiments proved him right. Albert Einstein, in deriving the photoelectric effect in 1905, put forward the theory that light was made of particles, just as Newton had argued. Einstein's light particle eventually became known as a *photon*, a name derived from the Greek word for light. According to the quantum theory, light may be both a wave and a particle, and this duality—and apparent paradox—is a mainstay of modern physics. Mysteriously, light exhibits *both* phenomena that are characteristic of waves, interference and diffraction, and phenomena of particles, localized in their interaction with matter. Two light rays interfere with each other in a way that is very similar to sound waves emanating from two stereo speakers, for example. On the other hand, light interacts with matter in a way that only particles can, as happens in the photoelectric effect.

Young's experiment showed that light was a wave. But we also know that light is, in a way, a particle: a photon. In the twentieth century, the Young experiment was repeated with very weak light—light that was produced as one photon at a time. Thus, it was very unlikely that several photons would be found within the experimental apparatus at the same time. Stunningly, *the same interference pattern appeared* as enough time elapsed so that the photons, arriving one at a time, accumulated on the screen. What was each photon interfering with, if it was *alone* in the experimental apparatus? The answer seemed to be: with itself. In a sense, each photon went through *both* slits, not one slit, and as it appeared on the other side, it interfered with itself.

The Young experiment has been carried out with many entities we consider to be particles: electrons, since the 1950s; neutrons, since the 1970s; and atoms, since the 1980s. In each case, the same interference pattern occurred. These findings demonstrated the de Broglie principle, according to which particles also exhibit wave phenomena. For example, in 1989, A. Tonomura and colleagues performed a double-slit experiment with electrons. Their results are shown below, clearly demonstrating an interference pattern.

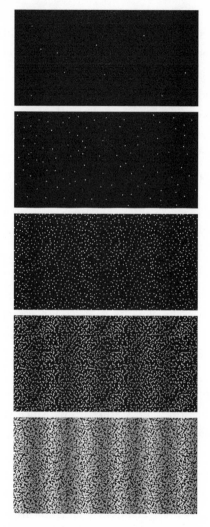

Anton Zeilinger and colleagues demonstrated the same pattern for neutrons, traveling at only 2 km/second, in 1991. Their results are shown below.

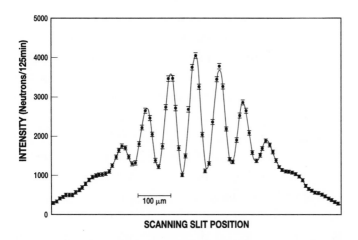

SCANNING SLIT POSITION

The same pattern was shown with atoms. This demonstrated that the duality between particles and waves manifests itself even for larger entities.

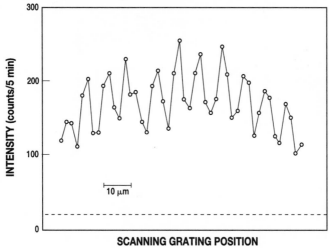

SCANNING GRATING POSITION

Anton Zeilinger and his colleagues at the University of Vienna, where Schrödinger and Mach had worked, went one step further. They extended our knowledge about quantum systems to entities that one would not necessarily associate any more with the world of the very small. (Although it should be pointed out that physicists know macroscopic systems, such as superconductors, that behave quantum-mechanically.) A *bucky ball* is a molecule of sixty or seventy atoms of carbon arranged in a structure resembling a geodesic dome. Buckminster Fuller made such domes famous, and the bucky ball is named after him. A molecule of sixty atoms is a relatively large entity, as compared with an atom. And yet, the same mysterious interference pattern appeared when Zeilinger and his colleagues ran their experiment. The arrangement is shown below.

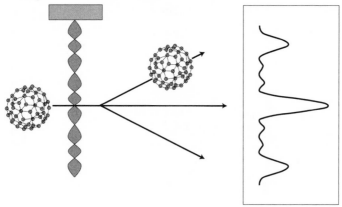

In each case, we see that the particles behave like waves. These experiments were also carried out one particle at a time, and still the interference pattern remained. What were these particles interfering with? The answer is that, in a sense, each particle went not through one slit, but rather through *both* slits—and then the particle "interfered with itself."

What we are witnessing here is a manifestation of the quantum *principle of superposition of states.*

The superposition principle says that a new state of a system may be composed from two or more states, in such a way that the new state shares some of the properties of each of the combined states. If A and B ascribe two different properties to a particle, such as being at two different places, then the *superposition* of states, written as A + B, has something in common both with state A and with state B. In particular, the particle will have non-zero probabilities for being in each of the two states, but not elsewhere, if the position of the particle is to be observed.

In the case of the double-slit experiment, the experimental setup provides the particle with a particular kind of superposition: The particle is in state A when it passes through slit A and in state B when it passes through slit B. The superposition of states is a combination of "particle goes through slit A" with "particle goes through slit B." This superposition of states is written as A + B. The two paths are combined, and there are therefore two nonzero probabilities, if the particle is observed. Given that the particle is to be observed as it goes through the experimental setup, it will have a 50% chance of being observed to go through slit A and a 50% chance that it will be observed to go through slit B. But if the particle is *not* observed as it goes through the experimental setup, only at the end as it collects on the screen, the superposition holds through to the end. In a sense, then, the particle has gone through both slits, and as it arrived at the end of the experimental setup, it interfered with itself. Superposition of states is the greatest mystery in quantum mechanics. The superposition principle encompasses within itself the idea of entanglement.

WHAT IS ENTANGLEMENT?

Entanglement is an application of the superposition principle to a composite system consisting of two (or more) subsystems. A subsystem here is a single particle. Let's see what it means when we say that the two particles are entangled. Suppose that particle 1 can be in one of two states, A or C, and that these states represent two contradictory properties, such as being at two different places. Particle 2, on the other hand, can be in one of two states, B or D. Again these states could represent contradictory properties such as being at two different places. The state AB is called a product state. When the entire system is in state AB, we know that particle 1 is in state A and particle 2 is in state B. Similarly, the state CD for the entire system means that particle 1 is in state C and particle 2 is in state D. Now consider the state AB + CD. We obtain this state by applying the superposition principle to the entire, two-particle system. The superposition principle allows the system to be in such a combination of states, and the state AB + CD for the entire system is called an *entangled state*. While the product state AB (and similarly CD) ascribes definite properties to particles 1 and 2 (meaning, for example, that particle 1 is in location A and particle 2 is in location B), the entangled state—since it constitutes a superposition—does not. It only says that there are possibilities concerning particles 1 and 2 that are correlated, in the sense that if measurements are made, then if particle 1 is found in state A, particle 2 must be in state B; and similarly if particle 1 is in state C, then particle 2 will be in state D. Roughly speaking, when particles 1 and 2 are entangled, there is no way to characterize either one of them by itself without referring to the other as well. This is so even though we can refer to each particle alone when the two are in the product state AB or CD, but not when they are in the superposition AB + CD. It is the superposition of the two product states that produces the entanglement.

4

Planck's Constant

"Planck had put forward a new, previously unimagined thought, the thought of the atomistic structure of energy."

—Albert Einstein

The quantum theory, with its bizarre consequences, was born in the year 1900, thirty-five years before Einstein and his colleagues raised their question about entanglement. The birth of the quantum theory is attributed to the work of a unique individual, Max Planck. Max Planck was born in Kiel, Germany, in 1858. He came from a long line of pastors, jurists, and scholars. His grandfather and great grandfather were both theology professors at the University of Göttingen. Planck's father, Wilhelm J. J. Planck, was a professor of Law in Kiel, and inspired in his son a deep sense of knowledge and learning. Max was his sixth child. Max's mother came from a long line of pastors. The family was wealthy and vacationed every year on the shores of the Baltic Sea and traveled through Italy and Austria. The family was liberal in its views and, unlike many

Germans of the time, opposed Bismarck's politics. Max Planck saw himself as even more liberal than his family.

As a student, Max was good but not excellent—he was never at the top of his class although his grades were generally satisfactory. He exhibited a talent for languages, history, music, and mathematics, but never cared much for, nor excelled in, physics. He was a conscientious student and worked hard, but did not exhibit great genius. Planck was a slow, methodical thinker, not one with quick answers. Once he started working on something he found it hard to leave the subject and move on to something else. He was more a plodder than a naturally gifted intellect at the gymnasium. He often said that, unfortunately, he had not been given the gift of reacting quickly to intellectual stimulation. And he was always surprised that others could pursue several lines of intellectual work. He was shy, but was always well-liked by his teachers and fellow students. He saw himself as a moral person, one loyal to duties, perfectly honest, and pure of conscience. A teacher at the gymnasium encouraged him to pursue the harmonious interplay that he thought existed between mathematics and the laws of nature. This prompted Max Planck to study physics, which he did upon entering the University of Munich.

In 1878, Planck chose thermodynamics as the topic for his dissertation, which he completed in 1879. The thesis dealt with two principles of classical thermodynamics: the conservation of energy, and the increase of entropy with time, which characterize all observable physical processes. Planck extracted some concrete results from the principles of thermodynamics and added an important premise: A stable equilibrium is obtained at a point of maximum entropy. He emphasized that thermodynamics can produce good results without any reliance whatever on the atomic hypothesis. Thus a system could be studied based on its macroscopic properties without the scientist having to worry about what happens or doesn't happen to the system's tiny components: atoms, molecules, electrons, and so on.

Thermodynamic principles are still extremely important in physics as they deal with the energy of entire systems. These principles can be used to determine the output of an internal combustion engine, for example, and have wide applicability in engineering and other areas. Energy and entropy are key concepts in physics, so one would have thought that Planck's work would have been well-received at the time. But it wasn't. Professors at Munich, and Berlin—where Planck had studied for a year—were not impressed by his work. They did not think the work was important enough to merit praise or recognition. One professor evaded Planck so he could not even serve him with a copy of his doctoral work when preparing for its defense. Eventually Planck was awarded the degree and was fortunate enough to obtain the position of associate professor at the University of Kiel, where his father still had a number of friends who could help him. He took his position in 1885 and immediately attempted to vindicate both his own work and thermodynamics as a whole. He entered a competition organized by the University of Göttingen to define the nature of energy. Planck's essay won second place—there was no first place. He soon realized that he would have had first place had not his article been critical of one of the professors at Göttingen.

Nonetheless, his award impressed the physics professors at the University of Berlin, and they offered him a post of associate professor in their faculty in 1889.

In time, the world of theoretical physics came to appreciate the principles of thermodynamics with their treatment of the concepts of energy and entropy, and Planck's work became more popular. His colleagues in Berlin, in fact, borrowed his dissertation so frequently that within a short time the manuscript started falling apart. In 1892 Planck was promoted to full professor in Berlin and in 1894 he became a full member of the Berlin Academy of Sciences.

By the late 1800s, physics was considered a completed discipline, within which all explanations for phenomena and experimental outcomes had already been satisfactorily given. There was mechanics, the theory started by Galileo with his reputed experiment of dropping items from atop the Leaning Tower of Pisa, and perfected by the genius of Isaac Newton by the turn of the eighteenth century, almost two centuries before Planck's time. Mechanics and the theory of gravitation that goes with it attempt to explain the motions of objects of the size we see in everyday life up to the size of planets and the moon. It explains how objects move; that force is the product of mass and acceleration; the idea that moving objects have inertia; and that the earth exerts a gravitational pull on all objects. Newton taught us that the moon's orbit around the earth is in fact a constant "falling" of the moon down to earth, impelled by the gravitational pull both masses exert on each other.

Physics also included the theory of electricity and electromagnetism developed by Ampere, Faraday, and Maxwell. This theory incorporated the idea of a field—a magnetic or electric field that cannot be seen or heard or felt, but which exerts its influence on objects. Maxwell developed equations that accurately described the electromagnetic field. He concluded that light waves are waves of the electromagnetic field. In 1831 Faraday constructed the first dynamo, which produced electricity through the principle of electromagnetic inductance. By rotating a copper disc between two poles of an electromagnet, he was able to produce current.

In 1887, during Planck's formative years, Heinrich Rudolf Hertz (1857-1894) conducted his experiments that produced radio waves. By chance, he noticed that a piece of zinc illuminated by ultraviolet light became electrically charged. Without knowing it, he had discovered the photoelectric effect, which links light with matter. Around the same time, Ludwig Boltzmann (1844-1906) assumed that gases consist of molecules and treated their behavior using statistical methods. In 1897, one of the most important discoveries of science took place: the existence of the electron was deduced by J. J. Thomson.

Energy was a crucial idea within all of these various parts of classical physics. In mechanics, half the mass times the velocity squared was defined as a measure of *kinetic* energy (from the Greek word *kinesis*, motion); there was another kind of energy, called *potential* energy. A rock on a high cliff possesses potential energy, which could then be instantly converted into kinetic energy once the rock is pushed slightly and falls from the cliff. Heat is energy, as we learn in high school physics. Entropy is a quality related to randomness and since randomness always increases, we have the law of increasing entropy—as everyone who has tried to put away toys knows well.

So there was every reason for the world of physics to accept Planck's modest contributions to the theories of energy and entropy, and this was indeed what happened in Germany toward the end of the nineteenth century. Planck was recognized for his work in thermodynamics, and became a professor at the University of Berlin. During that time, he started to work on an interesting problem. It had to do with what is known as black-body radiation. Logical reasoning along the lines of classical physics led to the conclusion that radiation from a hot object would be very bright at the blue or violet end of the spectrum. Thus a log in a fireplace, glowing red, would end up emitting ultraviolet rays as well as x-rays and gamma rays. But this phenomenon, known as the ultraviolet catastrophe, doesn't really take place in nature. No one knew how to explain this odd fact, since the theory did predict this buildup of energy levels of radiation. On December 14, 1900, Max Planck presented a paper at a meeting of the German Physical Society. Planck's conclusions were so puzzling that he himself found it hard to believe them. But these conclusions were the only logical explanation to the fact that the ultraviolet catastrophe does not occur. Planck's thesis was that energy levels are quantized. Energy does not grow or diminish continuously, but is rather always a multiple of a basic *quantum,* a quantity Planck denoted as hv, where v is a characteristic frequency of the system being considered, and h is a fundamental constant now known as Planck's constant. (The value of Planck's constant is 6.6262×10^{-34} joule-seconds.)

The Rayleigh-Jeans law of classical physics implied that the brightness of the black-body radiation would be unlimited at the extreme ultraviolet end of the spectrum, thus producing the ultraviolet catastrophe. But nature did not behave this way. According to nineteenth century physics (the work of Maxwell and Hertz), an oscillating charge produces radiation. The frequency (the inverse of the wavelength) of this oscillating charge is denoted by v, and its energy is E. Planck proposed a formula for the energy levels of a Maxwell-Hertz oscillator based on his constant h. The formula is:

$E=0$, hv, $2hv$, $3hv$, $4hv$ . . . , or in general, nhv, where n is a non-negative integer.

Planck's formula worked like magic. It managed to explain energy and radiation within a black body cavity in perfect agreement with the energy curves physicists were obtaining through their experiments. The reason for this was that the energy was now seen as coming in discrete packages, some large and some small, depending on the frequency of oscillation. But now, when the allotted energy for an oscillator (derived by other means) was *smaller* than the size of the package of energy available through Planck's formula, the intensity of the radiation dropped, rather than increasing to the high levels of the ultraviolet catastrophe.

Planck had invoked the quantum. From that moment on, physics was never the same. Over the following decades, many confirmations were obtained that the quantum is indeed a real concept, and that nature really works this way, at least in the micro-world of atoms, molecules, electrons, neutrons, photons and the like.

Planck himself remained somewhat baffled by his own discovery. It is possible that he never quite understood it on a philosophical level. The trick worked, and the equations fit the data, but the question: "Why the quantum?" was one that not only he, but generations of future physicists and philosophers would ask and continue asking.

Planck was a patriotic German who believed in German science. He was instrumental in bringing Albert Einstein to Berlin in 1914 and in promoting Einstein's election to the Prussian Academy of Sciences. When Hitler came to power, Planck tried to persuade him to change his decision to terminate the positions of Jewish academics. But Planck never quit his own position in protest, as some non-Jewish academics did. He remained in Germany, and throughout his life continued to believe in promoting science in his homeland.

Planck died in 1947. By that time, the quantum theory had matured and undergone significant growth to become the accepted theory of physical law in the world of the very small. Planck himself, whose work and discovery of quanta had initiated the revolution in science, never quite accepted it completely in his own mind. He seemed to be puzzled by the discoveries he had made, and at heart always remained a classical physicist, in the sense that he did not participate much in the scientific revolution that he had started. But the world of science moved forward with tremendous impetus.

5

The Copenhagen School

"The discovery of the quantum of action shows us not only the natural limitation of classical physics, but, by throwing a new light upon the old philosophical problem of the objective existence of phenomena independently of our observations, confronts us with a situation hitherto unknown in natural science."

—Niels Bohr

Niels Bohr was born in Copenhagen in 1885, in a six-teenth-century palace situated across the street from the Danish Parliament. The impressive building was owned by a succession of wealthy and famous people, including, two decades after Bohr's birth, King George I of Greece.

The palace was bought by David Adler, Niels's maternal grandfather, a banker and member of the Danish Parliament. Bohr's mother, Ellen Adler, came from an Anglo-Jewish family that had settled in Denmark. On his father's side, Niels belonged to a family that had lived in Denmark for many generations, emigrating there in the late 1700s from the Grand Duchy of Mecklenburg in the Danish-speaking part of Germany. Niels's father, Christian Bohr, was a physician and scientist who was nominated for the Nobel Prize for his research on respiration.

David Adler also owned a country estate about ten miles from Copenhagen, and Niels was raised in very comfortable surroundings both in the city and in the country. Niels attended school in Copenhagen and was nicknamed "the fat one," since he was a large boy who frequently wrestled with his friends. He was a good student, although not the first in his class.

Bohr's parents allowed their children to develop their gifts to the fullest. Bohr's younger brother, Harald, always showed a propensity for mathematics, and, in time, became a prominent mathematician. Niels stood out as a curious investigator even as a very young child. While still a student, Niels Bohr undertook a project to investigate the surface tension of water by observing the vibrations of a spout. The project was planned and executed so well that it won him a gold medal from the Danish Academy of Sciences.

At the university, Bohr was particularly influenced by Professor Christian Christiansen, who was the eminent Danish physicist of the time. The professor and the student had a relationship of mutual admiration. Bohr later wrote that he was especially fortunate to have come under the guidance of Christiansen, "a profoundly original and highly endowed physicist." Christiansen, in turn, wrote Bohr in 1916: "I have never met anybody like you who went to the bottom of everything and also had the energy to pursue it to completion, and who in addition was so interested in life as a whole."[4]

Bohr was also influenced by the work of the leading Danish philosopher, Harald Høffding. Bohr had known Høffding long before coming to the university, since he was a friend of

Bohr's father. Høffding and other Danish intellectuals regularly met at the Bohr mansion for discussions, and Christian Bohr allowed his two sons, Niels and Harald, to listen to the discussions. Høffding later became very interested in the philosophical implications of the quantum theory, developed by Niels Bohr. Some have suggested that, in turn, Bohr's formulation of the quantum principle of complementarity (discussed later) was influenced by the philosophy of Høffding. Bohr continued on to his Ph.D. in physics at the university, and in 1911 wrote his thesis on the electron theory of metals. In his model, metals are viewed as a gas of electrons moving more or less freely within the potential created by the positive charges in the metal. These positive charges are the nuclei of the atoms of the metal, arranged in a lattice. The theory could not explain everything, and its limitations were due to the application of classical—rather than the nascent *quantum*—ideas to the behavior of these electrons in a metal. His model worked so well that his dissertation defense attracted much attention and the room was full to capacity. Professor Christiansen presided over the proceedings. He remarked that it was unfortunate that the thesis had not been translated into a foreign language as well, since few Danes could understand the physics. Bohr later sent copies of his thesis to a number of leading physicists whose works he had made reference to in the thesis, including Max Planck. Unfortunately few responded, since none could understand Danish. In 1920, Bohr made an effort to translate the thesis into English, but never completed the project.

Having finished his work, Bohr went to England on a postdoctoral fellowship supported by the Danish Carlsberg

Foundation. He spent a year under the direction of J.J. Thomson at the Cavendish laboratory in Cambridge. The Cavendish laboratory was among the world's leading centers for experimental physics, and its directors before Thomson were Maxwell and Rayleigh. The laboratory has produced some twenty-odd Nobel laureates over the years.

Thomson, who had won the Nobel Prize in 1906 for his discovery of the electron, was very ambitious. Often the film taken during experiments had to be hidden from him so he wouldn't snatch it before it was dry to inspect it, leaving fingerprints that blurred the pictures. He was on a crusade to rewrite physics in terms of the electron, and to push beyond the impressive work of his predecessor, Maxwell.

Bohr worked hard in the laboratory, but often had difficulties blowing glass to make special equipment. He broke tubes, and fumbled in the unfamiliar language. He tried to improve his English by reading Dickens, using his dictionary for every other word. Additionally, Thomson was not easy to work with. The project Thomson assigned to Bohr had to do with cathode ray tubes, and was a dead end that did not yield any results. Bohr found an error in Thomson's calculations, but Thomson was not one who could accept criticism. He was uninterested in being corrected, and Bohr—with his poor English—did not make himself understood.

In Cambridge, Bohr met Lord James Rutherford (1871-1937), who was recognized for his pioneering work on radiation, the discovery of the nucleus, and a model of the atom. Bohr was interested in moving to Manchester to work with Rutherford, whose theories had not yet received widespread acceptance. Rutherford welcomed him but suggested that he first obtain Thomson's permission to leave. Thomson—who was not a believer in Rutherford's theory of the nucleus—was more than happy to let Bohr go.

In Manchester, Bohr began the studies that would eventually bring him fame. He started to analyze the properties of atoms in light of Rutherford's theory. Rutherford set Bohr to work on the experimental problem of analyzing the absorption of alpha particles in aluminum. Bohr worked in the lab many hours a day, and Rutherford visited him and the rest of his students often, showing much interest in their work. After a while, however, Bohr approached Rutherford and said that he would rather do theoretical physics. Rutherford agreed and Bohr stayed home, doing research with pencil and paper and rarely coming into the lab. He was happy not to have to see anyone, he later said, as "no one there knew much."

Bohr worked with electrons and alpha particles in his research, and produced a model to describe the phenomena that he and the experimental physicists were observing. The classical theory did not work, so Bohr took a big step: He applied quantum constraints to his particles. Bohr used Planck's constant in two ways in his famous theory of the hydrogen atom. First, he noted that the angular momentum of the orbiting electron in his model of the hydrogen atom had the same dimensions as Planck's constant. This led him to postulate that the angular momentum of the orbiting electron must be a multiple of Planck's constant divided by 2π, that is:

$$mvr = h/2\pi, \; 2(h/2\pi), \; 3(h/2\pi), \; \ldots$$

Where the expression on the left is the classical definition of angular momentum (m is mass, v is speed, and r is the radius of the orbit). This assumption of the quantizing of the angular momentum led Bohr directly to quantizing the energy of the atom.

Second, Bohr postulated that when the hydrogen atom drops from one energy level to a lower one, the energy that is released comes out as a single Einstein photon. As we will see later, the smallest quantity of energy in a light beam, according to Einstein, was hv, where h was Planck's constant and v the frequency, measured as the number of vibrations per second. With this development, and with his assumption of angular momentum, Bohr used Planck's quantum theory to explain what happens in the interior of an atom. This was a major breakthrough for physics.

Bohr finished his paper on alpha particles and the atom after he left Manchester and returned to Copenhagen. The paper was published in 1913, marking the transition of his work to the quantum theory and the question of atomic structure. Bohr never forgot he was led to formulate his quantum theory of the atom from Rutherford's discovery of the nucleus. He later described Rutherford as a second father to him.

Upon his return to Denmark, Bohr took up a position at the Danish Institute of Technology. He married Margrethe Nørlund in 1912. She remained by his side throughout his life, and was a power in organizing the physics group founded in Copenhagen by her husband.

On March 6, 1913, Bohr sent Rutherford the first chapter of his treatise on the constitution of atoms. He asked his former mentor to forward the work to the *Philosophical Magazine* for publication. This manuscript was to catapult him from a young physicist who has made some important progress in physics to a world figure in science. Bohr's breakthrough discovery was that it is impossible to describe the atom in classical terms, and that the answers to all questions about atomic phenomena had to come from the quantum theory.

Bohr's efforts were aimed at first understanding the simplest atom of all, that of hydrogen. By the time he addressed the problem, physics had already learned that there are specific series of frequencies at which the hydrogen atom radiates. These are the well-known series of Rydberg, Balmer, Lyman, Paschen, and Brackett—each covering a different part of the spectrum of radiation from excited hydrogen atoms, from ultraviolet through visible light to infrared. Bohr sought to find a formula that would explain why hydrogen radiates in these particular frequencies and not others.

Bohr deduced from the data available on all series of radiation of hydrogen that every emitted frequency was due to an electron descending from one energy level in the atom to another, lower level. When the electron came down from one level to another, the difference between its beginning and ending energies was emitted in the form of a *quantum of energy*. There is a formula linking these energy levels and quanta:

$$E_a - E_b = h\nu_{ab}$$

Where E_a is the beginning energy level of the electron around the hydrogen nucleus; E_b is the ending energy level once the electron has descended from its prior state; h is Planck's constant; and ν_{ab} is the frequency of the light quantum emitted during the electron's jump down from the first to the second energy level. This is demonstrated by the figure below.

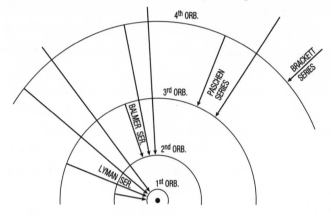

Rutherford's simple model of the atom did not square well with reality. Rutherford's atom was modeled according to classical physics, and if the atom was as simple as the model implied, it would not have existed for more than one-hundred-millionth of a second. Bohr's tremendous discovery of the use of Planck's constant within the framework of the atom solved the problem beautifully. The quantum theory now explained all observed radiation phenomena about hydrogen, which had until then baffled physicists for decades.

Bohr's work has been partially extended to explain the orbits and energies of electrons in other elements and to bring us understanding of the periodic table of the elements, chemical bonds, and other fundamental phenomena. The quantum theory had just been put to exceptionally good use. It was becoming obvious that classical physics would not work well in the realm of atoms and molecules and electrons, and that the quantum theory was the correct path to take.

Bohr's brilliant solution to the question of the various series of spectral lines of radiation for the hydrogen atom left unanswered the question: Why? Why does an electron jump from one energy level to another, and how does the electron know that it should do so? This is a question of causality. Causality is not explained by the quantum theory, and in fact cause and effect are blurred in the quantum world and have no explanation or meaning. This question about Bohr's work was raised by Rutherford as soon as he received Bohr's manuscript. Also, the discoveries did not bring about a *general formulation* of quantum physics, applicable in principle to all situations and not just to special cases. This was the main question of the time, and the goal was not achieved until later, that is, until the birth of "the new quantum mechanics" with the work of de Broglie, Heisenberg, Schrödinger, and others.

Bohr became very famous following his work on the quantum nature of the atom. He petitioned the Danish government to endow him with a chair of theoretical physics, and the government complied. Bohr was now Denmark's favorite son and the whole country honored him. Over the next few years he continued to travel to Manchester to work with Rutherford, and traveled to other locations and met many physicists. These connections allowed him to found his own institute.

In 1918, Bohr secured permission from the Danish government to found his institute of theoretical physics. He received funding from the Royal Danish Academy of Science, which draws support from the Carlsberg brewery. Bohr and his family moved into the mansion owned by the Carlsberg family on the premises of his new institute. Many young physicists from around the world regularly came to spend a year or two working at the institute and drawing their inspiration from the great Danish physicist. Bohr became close with the Danish royal family, as well as with many members of the nobility and the international elite. In 1922, he received the Nobel Prize for his work on the quantum theory.

Bohr organized regular scientific meetings at his institute in Copenhagen, to which many of the world's greatest physicists came and discussed their ideas. Copenhagen thus became a world center for the study of quantum mechanics during the period the theory was growing: from its founding in the late first decade of the twentieth century until just before the Second World War. The scientists who worked at the institute (to be named the Niels Bohr Institute after its founder's death), and many who came to attend its meetings, later developed what is called the *Copenhagen Interpretation* of the quantum theory, often called the orthodox interpretation. This was done after the birth of the "new quantum mechanics" in the middle 1920s. According to the Copenhagen interpretation of the rules of the quantum world, there is a clear distinction between what is observed and what is not observed. The quantum system is submicroscopic and does not include the measuring devices or the measuring process. In the years to come, the Copenhagen interpretation would be challenged by newer views of the world brought about by the maturing of the quantum theory.

Starting in the 1920s, and culminating in 1935, a major debate would rage within the community of quantum physicists. The challenge would be issued by Einstein, and throughout the rest of his life, Bohr would regularly spar with Einstein on the meaning and completeness of the quantum theory.

6

De Broglie's Pilot Waves

"After long reflection in solitude and meditation, I suddenly
had the idea, during the year 1923, that the discovery made
by Einstein in 1905 should be generalized by extending it
to all material particles and notably to electrons."
                                                    —Louis de Broglie

Duke Louis Victor de Broglie was born in Dieppe in
1892 to an aristocratic French family that had long
provided France with diplomats, politicians, and
military leaders. Louis was the youngest of five children. His
family expected Louis' adored older bother, Maurice, to enter
the military service, and so Louis too decided to serve France.
He chose the navy, since he thought it might allow him to
study the natural sciences, which had fascinated him since
childhood. He did indeed get to practice science by installing
the first French wireless transmitter aboard a ship.

After Maurice left the military and studied in Toulon and
at the University of Marseilles, he moved to a mansion in
Paris, where in one of the rooms he established a laboratory
for the study of X-rays. To aid him in his experiments, the
resourceful Maurice trained his valet in the rudiments of sci-

entific procedure, and eventually converted his personal servant into a professional lab assistant. His fascination with science was infectious. Soon, his younger brother Louis was also interested in the research and helped him with experiments.

Louis attended the Sorbonne, studying medieval history. In 1911, Maurice served as the secretary of the famous Solvay Conference in Brussels, where Einstein and other leading physicists met to discuss the exciting new discoveries in physics. Upon his return, he regaled his younger brother with stories about the fascinating discoveries, and Louis became even more excited about physics.

Soon, World War I erupted and Louis de Broglie enlisted in the French army. He served in a radio communication unit, a novelty at that time. During his service with the radio-telegraphy unit stationed at the top of the Eiffel Tower, he learned much about radio waves. And indeed he was to make his mark on the world through the study of waves. When the war ended, de Broglie returned to the university and studied under some of France's best physicists and mathematicians, including Paul Langevin and Emile Borel. He designed experiments on waves and tested them out at his brother's laboratory in the family's mansion. De Broglie was also a lover of chamber music, and so he had an intimate knowledge of waves from a music-theory point of view.

De Broglie immersed himself in the study of the proceedings of the Solvay Conference given to him by his brother. He was taken by the nascent quantum theory discussed in 1911 and repeatedly presented at later Solvay meetings throughout the following years. De Broglie studied ideal gases, which were discussed at the Solvay meeting, and came to a successful implementation of the theory of waves in analyzing the physics of such gases, using the quantum theory.

In 1923, while working for a doctorate in physics in Paris, "all of a sudden," as he later put it, "I saw that the crisis in optics was simply due to a failure to understand the true universal duality of wave and particle." At that moment, in fact, de Broglie discovered this duality. He published three short notes on the topic, hypothesizing that particles were also waves and waves also particles, in the *Proceedings* of the Paris Academy in September and October 1923. He elaborated on this work and presented his entire discovery in his doctoral dissertation, which he defended on November 25, 1924.

De Broglie took Bohr's conception of an atom and viewed it as a musical instrument that can emit a basic tone and a sequence of overtones. He suggested that all particles have this kind of wave-aspect to them. He later described his efforts: "I wished to represent to myself the union of waves and particles in a concrete fashion, the particle being a little localized object incorporated in the structure of a propagating wave." Waves that he associated with particles, de Broglie named pilot waves. Every small particle in the universe is thus associated with a wave propagating through space.

De Broglie derived some mathematical concepts for his pilot waves. Through a derivation using several formulas and Planck's quantum-theory constant, h, de Broglie came up with the equation that is his legacy to science. His equation links the momentum of a particle, p, with the wavelength of its associated pilot wave, λ, through an equation using Planck's constant. The relationship is very simply stated as:

$$p=h/\lambda$$

De Broglie had a brilliant idea. Here, he was using the machinery of the quantum theory to state a very explicit relationship between particles and waves. A particle has momentum (classically, the product of its velocity and its mass). Now this momentum was directly linked with the wave associated with the particle. Thus a particle's momentum in quantum mechanics is, by de Broglie's formula, equal to the quotient of Planck's constant and the wavelength of the wave when we view the particle as a wave.

De Broglie did not provide an equation to describe the propagation of the wave associated with a particle. That task would be left to another great mind, Erwin Schrödinger. For his pioneering work, de Broglie received the Nobel Prize after many experiments verified the wave nature of particles over the following years.

De Broglie remained active as a physicist and lived a long life, dying in 1987 at the age of 95. When de Broglie was already a world-famous scientist, the physicist George Gamow (who wrote *Thirty Years that Shook Physics*) visited him in his mansion in Paris. Gamow rang the bell at the gate of the estate and was greeted by de Broglie's butler. He said: "Je voudrais voir Professeur de Broglie." The butler cringed. "Vous voulez dire, Monsieur le Duc de Broglie!" he insisted. "O.K., le Duc de Broglie," Gamow said and was finally allowed to enter.

* * *

Are particles also waves? Are waves also particles? The answer the quantum theory gives us is "Yes." A key characteristic of a quantum system is that a particle is also a wave, and exhibits wave interference characteristics when passed through a double-slit experimental setup. Similarly, waves can be particles, as Einstein has taught us when he developed his photoelectric effect Nobel Prize-winning paper, which will be described later. Light waves are also particles, called photons. Laser light is coherent light, in which all the light waves are in phase; hence the power of lasers. In 2001, the Nobel Prize in physics was shared by three scientists who showed that atoms, too, can behave like light rays in the sense that an ensemble of them can all be in a coherent state, just like laser light. This proved a conjecture put forth by Einstein and his colleague, the Indian physicist Saryendra Nath Bose, in the 1920s. Bose was an unknown professor of physics at the University of Dacca, and in 1924 he wrote Einstein a letter in which he described how Einstein's light quanta, the photons, could form a kind of "gas," similar to the one consisting of atoms or molecules. Einstein rewrote and improved Bose's paper and submitted it for joint publication. This gas proposed by Bose and Einstein was a new form of matter, in which individual particles did not have any properties and were not distinguishable. The Bose-Einstein new form of matter led Einstein to a "hypothesis about an interaction between molecules of an as yet quite mysterious nature."

The Bose-Einstein statistics allowed Einstein to make groundbreaking predictions about the behavior of matter at extremely low temperatures. At such low temperatures, viscosity of liquefied gases disappears, resulting in superfluidity. The process is called Bose-Einstein condensation.

Louis de Broglie had submitted his doctoral dissertation to Einstein's friend in Paris, Paul Langevin, in 1924. Langevin was so impressed with de Broglie's idea that matter can have a wave aspect, that he sent the thesis to Einstein, asking for his opinion. When Einstein read de Broglie's thesis he called it "very remarkable," and he later used the de Broglie wave idea to deduce the wave properties of the new form of matter he and Bose had discovered. But no one had seen a Bose-Einatein condensate . . . until 1995.

On June 5, 1995, Carl Weiman of the University of Colorado and Eric Cornell of the National Institute of Standards and Technology used high-powered lasers and a new technique for cooling matter to close to absolute zero to supercool about 2000 atoms of rubidium. These atoms were found to possess the qualities of a Bose-Einstein condensate. They appeared as a tiny dark cloud, in which the atoms themselves had lost their individuality and entered a single energy state. For all purposes, these atoms were now one quantum entity, as characterized by their de Broglie wave. Shortly afterwards, Wolfgang Ketterle of M.I.T. reproduced the results and improved the experiment, producing what was the equivalent of a laser beam made of atoms. For their work, the three scientists shared the 2001 Nobel Prize in Physics, and de Broglie's fascinating idea was reconfirmed in a new setting that pushed the limits of quantum mechanics up the scale toward macroscopic objects.

7

Schrödinger and His Equation

"Entanglement is not *one* but rather *the* characteristic trait of quantum mechanics."

—Erwin Schrödinger

*E*rwin Schrödinger was born in a house in the center of Vienna in 1887 to well-to-do parents. An only child, he was doted on by several aunts, one of whom taught him to speak and read English before he even mastered his native German. As a young boy, Erwin started to keep a journal, a practice he maintained throughout his life. From an early age, he exhibited a healthy skepticism and tended to question things that people presented as facts. These two habits were very useful in the life of a scientist who would make one of the most important contributions to the new quantum theory. Questioning what from our everyday life we take as truth is essential in approaching the world of the very small. And Schrödinger's notebooks would be crucial in his development of the wave equation.

At age eleven, Erwin entered the gymnasium located a few

minutes' walk from his house. In addition to mathematics and the sciences, the gymnasium taught its students Greek language and culture, Latin, and the classic works of antiquity, including Ovid, Livy, Cicero, and Homer. Erwin loved mathematics and physics, and excelled in them, solving problems with an ease and facility that stunned his peers. But he also enjoyed German poetry and the logic of grammar, both ancient and modern. This logic, in mathematics and in humanistic studies, shaped his thinking and prepared him for the rigors of the university.

Erwin loved hiking, mountaineering, the theater, and pretty girls—amusements that would mark his behavior throughout his life. As a young boy, he worked hard at school, but also played hard. He spent many days walking in the mountains, reading mathematics, and courting his best friend's sister, a dark-haired beauty named Lotte Rella.[5]

In 1906, Schrödinger enrolled at the University of Vienna—one of the oldest in Europe, established in 1365— to study physics. There was a long legacy of physics at the university. Some of the great minds that had worked there and left about the time Schrödinger enrolled were Ludwig Boltzmann, the proponent of the atomic theory, and Ernst Mach, the theoretician whose work inspired Einstein. There Schrödinger was a student of Franz Exner, and did work in experimental physics, some of it relating to radioactivity. The University of Vienna was an important center for the study of radioactivity, and Marie Curie in Paris received some of her specimens of radioactive material, with which she made her discoveries, from the physics department at Vienna.

Schrödinger was admired by his fellow students for his

brilliance in physics and mathematics. He was always sought out by his friends for help in mathematics. One of the subjects in mathematics that he took at the University of Vienna was differential equations, in which he excelled. As fate would have it, this special skill proved invaluable in his career: it helped him solve the biggest problem of his life and establish his name as a pioneer of quantum mechanics.

But Schrödinger lived a multifacted life as a university student in Vienna at the height of its imperial glory. He retained his abilities as an athlete and was as highly social as he'd ever been: He found a number of good friends with whom he spent his free time climbing and hiking in the mountains. Once, in the Alps, he spent an entire night nursing a friend who had broken a leg while climbing. Once his friend was taken to the hospital, he spent the day skiing.

In 1910, Schrödinger wrote his doctoral thesis in physics, entitled "On the conduction of electricity on the surface of insulators in moist air." This was a problem that had some implications in the study of radioactivity, but the thesis was not a brilliant work of scholarship. Schrödinger had left out a number of factors about which he should have known, and his analysis was neither complete nor ingenious. Still, the work was enough to earn him his doctorate, and following his graduation he spent a year in the mountains as a volunteer in the fortress artillery. He then returned to the university to work as an assistant in a physics laboratory. Meanwhile, he labored on the required paper (called a *Habilitationschrift*) that would allow him to earn income as a private tutor at the university. His paper, "On the Kinetic Theory of Magnetism," was a theoretical attempt to explain the magnetic properties of various compounds, and was also not of exceptional quality, but it satisfied the requirements and allowed him to work at the university. His academic career had begun.

Shortly afterwards, Schrödinger, who was now in his early twenties, met another teenage girl who caught his fancy. Her name was Felicie Krauss, and her family belonged to Austria's lower nobility. The two developed a relationship and considered themselves engaged despite strong objections from the girl's parents. Felicie's mother, especially, was determined not to allow her daughter to marry a working-class person; one who, she believed, would never be able to support her daughter in an appropriate style on his university income. In despair, Erwin contemplated leaving the university and working for his father, who owned a factory. But the father would hear nothing of it, and with the mounting pressure from Felicie's mother, the two lovers called off their informal engagement. While she later married, Felicie always remained close to Erwin. This, too, was a pattern that continued throughout Schrödinger's life: wherever he went—even after he was married—there were always young girlfriends never too far away.

Schrödinger continued his study of radioactivity in the laboratory of the University of Vienna. In 1912, his colleague Victor Hess soared 16,000 feet in a balloon with instruments to measure radiation. He wanted to solve the problem of why radiation was detected not only close to the ground, where deposits of radium and uranium were its source, but also in the air. Up in his balloon, Hess discovered to his surprise that the radiation was actually three times as high as it was at ground level. Hess had thus discovered cosmic radiation, for which he later received the Nobel Prize. Schrödinger, taking part in related experiments on the background radiation at ground level, traveled throughout Austria with his own radiation-detecting instruments. This travel incidentally allowed him to enjoy his beloved outdoors—and make new friends. In 1913, he was taking radiation measurements in the open air in the area where a family he had known from Vienna was vacationing. With the family was a pretty teenage girl, Annemarie ("Anny") Bertel. The twenty-six year old scientist and the sixteen-year-old girl were attracted to each other, and through meetings over the next several years, developed a romance that resulted in marriage. Anny remained devoted to Schrödinger throughout his life, even tolerating his perpetual relationships with other women.

In 1914, Schrödinger reenlisted in the fortress artillery to fight on the Italian front of World War I. Even in the field, he continued to work on problems of physics, publishing papers in professional journals. None of his papers thus far had been exceptional, but the topics were interesting. Schrödinger spent much time doing research on color theory, and made contributions to our understanding of light of different wavelengths. During one of his experiments on color while still at the University of Vienna, Erwin discovered that his own color vision was deficient.

In 1917, Schrödinger wrote his first paper on the quantum theory, on atomic and molecular heats. The research for this paper brought to his attention the work of Bohr, Planck, and Einstein. By the time the war was over, Schrödinger had addressed not only the quantum theory, but also Einstein's theory of relativity. He had now brought himself into the leading edge of theoretical physics.

In the years following the war, Schrödinger taught at universities in Vienna, Jena, Breslau, Stuttgart, and Zurich. In 1920, in Vienna, Erwin married Anny Bertel. Her income was higher than his university salary, which made him upset and prompted him to seek employment at other universities throughout Europe. Through Anny, Erwin met Hansi Bauer, who later became one of the girlfriends he would maintain throughout his life.

In Stuttgart, in 1921, Schrödinger began a very serious effort to understand and further develop the quantum theory. Bohr and Einstein, who were not much older than Schrödinger, had already made their contributions to the theory while in their twenties. Schrödinger was getting older, and he still had not had a major scientific achievement. He concentrated his efforts on modeling the spectral lines of alkali metals.

In late 1921, Schrödinger was nominated for a coveted position of full Professor of Theoretical Physics at the University of Zurich. That year, he published his first important paper in the quantum area, about quantized orbits of a single electron, based on the earlier work of Bohr. Soon after his arrival in Zurich, however, he was diagnosed with pulmonary disease and his doctors ordered rest at high altitude. The Schrödingers decided on the village of Arosa in the Alps, not far from Davos, at an altitude of 6,000 feet. Upon his recovery, they returned to Zurich and there, in 1922, Schrödinger gave his inaugural lecture at the university. During 1923 and 1924, Schrödinger's research was centered on spectral theory, light, atomic theory, and the periodic nature of the elements. In 1924, at the age of 37, he was invited to attend the Solvay Conference in Brussels, where the greatest minds in physics, including Einstein and Bohr, met. Schrödinger was there almost as an outside observer, since he had not produced any earth-shattering papers.

The quantum theory was nowhere near being complete, and Erwin Schrödinger was desperately seeking a topic in the quantum field with which he could make his mark. Time was running out on him, and if nothing happened soon, he would be condemned to obscurity, mediocrity, and to remain forever in the sidelines while others were making scientific history. In 1924, Peter Debye at the University of Zurich asked Schrödinger to report on de Broglie's thesis on the wave theory of particles at a seminar held at the university. Schrödinger read the paper, started thinking about its ideas, and decided to pursue them further. He worked on de Broglie's particle-wave notion for a full year, but made no breakthrough.

A few days before Christmas, 1925, Erwin left for the Alps, to stay in the Villa Herwig in Arosa, where he and Anny had spent several months while he was recuperating four years earlier. This time he came without his wife. From his correspondence, we know that he had one of his former girlfriends from Vienna join him at the villa, and stay with him there till early 1926. Schrödinger's biographer Walter Moore makes much of the mystery as to who the girlfriend might have been.[6] Could she have been Lotte, Felicie, Hansi, or one of his other liaisons? At any rate, according to the physicist Hermann Weyl, Schrödinger's erotic encounters with the mystery lady produced the burst of energy Schrödinger required to make his great breakthrough in the quantum theory. Over the Christmas vacation in the Alps with his secret lover, Schrödinger produced the now-famous *Schrödinger equation*. The Schrödinger equation is the mathematical rule that describes the statistical behavior of particles in the micro-world of quantum mechanics. The Schrödinger equation is a differential equation.

Differential equations are mathematical equations that state a relationship between a quantity and its derivatives, that is, between a quantity and its rate of change. Velocity, for example, is the derivative (the rate of change) of location. If you are moving at sixty miles per hour, then your location on the road changes at a rate of sixty m.p.h. Acceleration is the rate of change of velocity (when you accelerate, you are increasing the speed of your car); thus acceleration is the *second derivative* of location, since it is the rate of change of the rate of change of location. An equation that states your location, as a variable, as well as your velocity, is a differential equation. An equation relating your location with your velocity and your acceleration is a second-order differential equation.

By the time Schrödinger started to address the problem of deriving the equation that governs the quantum behavior of a small particle such as the electron, a number of differential equations of classical physics were known. For example, the equation that governs the progression of heat in a metal was known. Equations governing classical waves, for example, waves on a vibrating string, and sound waves, were already well known. Having taken courses in differential equations,

Schrödinger was well aware of these developments. Schrödinger's task was to find an equation that would describe the progression of particle waves, the waves that de Broglie had associated with small particles. Schrödinger made some educated guesses about the form his equation must take, based on the known classical wave equation. What he had to determine, however, was whether to use the first or the second derivative of the wave with respect to location, and whether to use the first or the second derivative with respect to time. His breakthrough occurred when he discovered that the proper equation is first-order with respect to time and second-order with respect to location.

$$H\Psi=E\Psi$$

The above is the time-independent Schrödinger equation, stated in its simplest symbolic way. The symbol Ψ represents the *wave function* of a particle. This is de Broglie's "pilot wave" of a particle. But here this is no longer some hypothetical entity, but rather a function that we can actually study and analyze using the Schrödinger equation. The symbol H stands for an *operator*, which is represented by a formula of its own, telling it what to do to the wave function: take a derivative and also multiply the wave function by some numbers, including Planck's constant, h. The operator H *operates* on the wave equation, and the result, on the other side of the equation, is an energy level, E, multiplied by the wave function.

Schrödinger's equation has been applied very successfully to a number of situations in quantum physics. What a physicist does is to write the equation above, applied to a particular situation, say, a particle placed in a microscopic box, or an electron placed in a potential field, or the hydrogen atom. In each situation, the physicist then *solves* Schrödinger's equation, obtaining a solution. The solutions of the Schrödinger equation are *waves*.

Waves are usually represented in physics by trigonometric functions, most often the sine and the cosine functions, whose graphs look just like the picture of a wave. (Physicists also use other functions, such as exponentials.) The picture below is a typical sine wave.

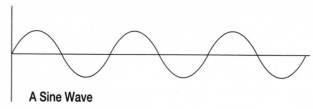

A Sine Wave

In solving the Schrödinger equation, a physicist will get a solution that states the wave function as something like: $\Psi = A \sin (n\pi x/L)$. (This is the solution for a particle in a rigid box. The term "sin" stands for the wave-like sine function, while all the other letters in the equation are constants or a variable (x). But the essential element here is the sine function.)

With his wave equation, Schrödinger brought quantum mechanics to a very high level. Scientists could now deal with a concrete wave function, which they could sometimes write down in specific terms, as in the example above, to describe particles or photons. This brought quantum theory to a point where several of its most important aspects are evident. Two of these notions are *probability* and *superposition*.

When we deal with quantum systems—each with an associated wave function Ψ—we no longer deal with precisely known elements. A quantum particle can only be described by its probabilities—never by exact terms. These probabilities are completely determined by the wave function, Ψ. The probabilistic interpretation of quantum mechanics was suggested by Max Born, although Einstein knew it first. The probability that a particle will be found in a given place is equal to the square of the amplitude of the wave function at that location:

$$\text{Probability} = |\Psi|^2$$

This is an extremely important formula in quantum theory. In many ways, it represents the essence of what quantum theory can give us. In classical physics, we can—in principle—measure, determine, and predict the location and the speed of a moving object with 100% certainty. This feature of classical (large-scale) physics is what allows us, for example, to land a spacecraft on the moon, not to mention drive a car or answer the door. In the world of the very small, we do not have these abilities to predict movements of particular objects. Our predictions are only statistical in their nature. We can determine where a particle will be (if the position observable is actualized) only in terms of probabilities of different outcomes (or, equivalently, what proportion of a large number of particles will fall at a specific location). Schrödinger's equation allows us to make such probabilistic predictions. As would be proven mathematically within a few decades, probabilities are all that we can get from quantum mechanics. There are no hidden quantities here whose knowledge would reduce the uncertainty. By its very nature, the quantum theory is probabilistic.

Probabilities are given by a probability distribution, which in the case of quantum theory is specified by the square of the amplitude of the wave function. Predicting the outcomes of quantum events is different from predicting the motion of a car, for example, where if you know the speed and initial location of the car, you will know its location after a certain period of time driving at a given speed, when both time and speed can be measured to great accuracy. If you drive for two hours at 60 mph, you will be 120 miles away from where you started. In the quantum world this does not happen. The best you can do is to predict outcomes in terms of probabilities. The situation, therefore, is similar to rolling two dice. Each die has a probability of 1/6 of landing on any given number. The two dice are independent, so the probability of rolling two sixes is the product of the probability of rolling a six on one die, 1/6, and the probability of rolling a six on the other die, again 1/6. The probability of two sixes is therefore 1/36. The probability of getting a sum of 12 on two dice is thus 1/36. The sum of two dice has the highest probability of equaling 7. That probability is 1/6. The distribution of the sum of two dice is shown below.

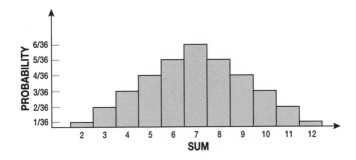

The square of the amplitude of the wave function, Ψ, is often a bell-shaped distribution. One such distribution is shown below.

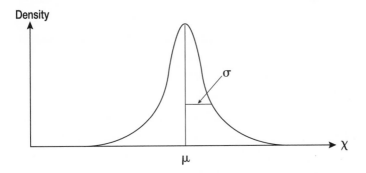

The distribution above gives us the probability of finding the particle in any given range of values of the horizontal axis by the area under the curve above that region.

The second essential element of the quantum theory brought to light by Schrödinger's equation is the *superposition principle*. Waves can always be superposed on one another. The reason for this is that the sine curve and the cosine curve for various parameters can be added to one another. This is the principle of Fourier analysis, discovered by the great French mathematician Joseph B. J. Fourier (1768-1830) and published in his book, *Théorie analytique de la chaleur* (Analytic Theory of Heat) in 1822. Fourier applied his theory to the propagation of heat, as suggested by the title of his book. He proved that many mathematical functions can be seen as the sum of many sine and cosine wave functions.

In quantum mechanics, because the solutions of Schrödinger's equation are waves, sums of these waves are also solutions to the equation. (The sum of several solutions of the Schrödinger equation is a solution because of the property of linearity.) This suggests, for example, that the electron can also be found in a state that is a superposition of other states. This happens because a solution of Schrödinger's equation for the electron would be some sine wave and thus a sum of such sine waves could also be a solution.

The superposition of waves explains the phenomenon of interference. In Young's double-slit experiment, the waves interfere with each other: that is, the bright lines on the screen are regions where the waves from the two slits add up to reinforce each other, while the dark streaks are regions where they subtract from each other, making the light weaker or completely absent.

Superposition is one of the most important principles in quantum mechanics. The weirdness of quantum mechanics really kicks in when a particle is superposed with *itself*. In Young's experiment, when the light is so weak that only one photon it emitted at a time, we still find the interference pattern on the screen. (The pattern is produced by many photons, not one, even though only one photon arrives at a time.)

The explanation for this phenomenon is that the single photon does not choose one slit *or* the other to go through. It chooses both slits, that is, one slit *and* the other. The particle goes through both slits, and then it interferes with itself, as two waves do by superposition.

When the quantum system contains more than one particle, the superposition principle gives rise to the phenomenon of *entanglement*. It is now not just a particle interfering with itself—it is a system interfering with itself: an entangled system. Amazingly enough, Erwin Schrödinger himself realized that particles or photons produced in a process that links them together will be entangled, and he actually coined the term entanglement, both in his native German and in English. Schrödinger discovered the possibility of entanglement in 1926, when he did his pioneering work on the new quantum mechanics, but he first used the term *entanglement* in 1935, in his discussion of the Einstein, Podolsky, and Rosen (EPR) paper.

According to Horne, Shimony, and Zeilinger, Schrödinger recognized in a series of papers in 1926 that the quantum state of an *n*-particle system can be entangled.[7] Schrödinger wrote:

> We have repeatedly drawn attention to the fact that the Ψ function itself cannot and may not be interpreted directly in terms of three-dimensional space—however much the one-electron problem tends to mislead us on this point—because it is in general a function in configuration space, not real space.[8]

According to Horne, Shimony, and Zeilinger, Schrödinger thus understood that the wave function in configuration space cannot be factored, which is a characteristic of entanglement. Nine years later, in 1935, Schrödinger actually named the phenomenon *entanglement*. He defined entanglement as follows:

> When two systems, of which we know the states by their respective representation, enter into a temporary physical interaction due to known forces between them and when after a time of mutual influence the systems separate again, then they can no longer be described as before, viz., by endowing each of them with a representative of its own. I would not call that *one* but rather *the* characteristic trait of quantum mechanics.[9]

In 1927, Schrödinger was appointed to succeed Max Planck as a Professor at the University of Berlin, and in 1929 he was further elected to membership in the Prussian Academy of Sciences. Then in May 1933, he gave up his position in disgust after Hitler was elected Chancellor of Germany, and exiled himself to Oxford. In 1933, Schrödinger was awarded the Nobel Prize for his great achievements in physics. He shared the award with the English physicist Paul Dirac, who also made important contributions to the quantum theory and predicted the existence of antimatter based on purely theoretical considerations.

Schrödinger returned to Austria and took up a professorship at the University of Graz. But when the Nazis took over Austria in 1938, he again escaped to Oxford. He did return to the Continent for one year and taught at Ghent, but as the War intensified he left for Dublin, where he became a professor of theoretical physics and held that position until 1956. While living in exile in Ireland, in the spring of 1944,

Schrödinger became involved in another extramarital affair. He was then 57 years old, and he got entangled with a young married woman, Sheila May Greene. He wrote her poetry, watched her perform in plays, and fathered her baby girl. Anny offered to divorce him so he could marry Sheila, but Erwin refused. The affair ended and David, Sheila's husband, raised the girl despite the fact that he and Sheila later separated. In 1956, Erwin finally returned to Vienna. He died there in 1961, his wife, Anny, by his side.

8

Heisenberg's Microscope

"To get into the spirit of the quantum theory was, I would say, only possible in Copenhagen at that time."
—Werner Heisenberg

Werner Carl Heisenberg (1901-1976) was born outside Munich, in southern Germany, and when he was still a young child, his family moved into the city. Throughout his life, Heisenberg felt at home in Munich and returned to it again and again from wherever he was living. At his sixtieth birthday celebration, organized by the city, Heisenberg said: "Anyone who didn't experience Munich in the Twenties has no idea how splendid life can be." His father, August Heisenberg, was a professor of Greek philology at the University of Munich, and in fact was the only full professor of middle and modern Greek philosophy in Germany. The father imparted to Werner a love of Greek ideas, and Heisenberg never lost his love of Plato. (Ironically, the ancient Greek concepts of time and space and causality would come into conflict with new notions brought on by the quantum theory created by Heisenberg and his colleagues.) While he was still in school, Heisenberg became interested in physics and decided to pursue a career as a scientist. He attended the University of Munich, and after finishing his undergraduate studies remained to study for a Ph.D. in physics.

In 1922, while a graduate student at Munich, Werner heard a public lecture on campus by Niels Bohr. He raised his hand and asked Bohr a hard question. When the lecture was over, Bohr came over to him and asked him to take a stroll. They walked for three hours, discussing physics. This was the beginning of a lifelong friendship.

After finishing his studies, Heisenberg came to attend Bohr's insitute in Copenhagen and spent the years 1924-1927 there, learning both Danish and English while pursuing his other studies. By 1924, when he was 23, Heisenberg had already written twelve papers on quantum mechanics, several of them cowritten with the great physicists Max Born and Arnold Sommerfeld. Heisenberg became Bohr's favorite disciple, and he often visited Bohr and his wife Margrethe at their home. When the big debate started between Einstein and Bohr, Heisenberg typically took Bohr's point of view, while Schrödinger sided with Einstein. This entanglement between Bohr and Heisenberg lasted throughout their lives.

Heisenberg developed a theory of quantum mechanics equivalent to Schrödinger's. His version was finished a little before that of his senior colleague. While Schrödinger's approach used his wave equation, Heisenberg's solution was based on matrices, conceptually more challenging. Matrix mechanics uses numbers in rows and columns to predict the intensities of the light waves emitted from "excited" atoms changing energy levels as well as other quantum phenomena.

It was later shown that the two methods are equivalent. In Heisenberg's more abstract approach, infinite matrices represent properties of observable entities, and the mathematics used is the mathematics of matrix manipulations. Matrix multiplication is non-commutative, meaning that if we multiply two matrices, A and B in the order AB, the answer, in general, will not be the same as the one we would get by performing the operation in reverse order, i.e., as BA. Contrast this with the way we multiply numbers, which is commutative (for example, 5×7=35=7×5, so that the order of multiplication does not matter and we get the same answer in both ways). The noncommutativity of the multiplication operation on matrices has important consequences in quantum mechanics, which go beyond the work of Heisenberg.

An *observable* (something about a quantum system that we may observe) is represented in modern quantum mechanics by the action of an *operator* on the wave function of the system. Some of these operators are commutative, meaning that if we apply one operator and then another to the system in the order AB, then the answer is the same as it would be if we applied the two operators in the reverse order: BA. Other operators are noncommutative, meaning that the order of applying the operators (and thus the order of making the observations) matters and the results are different from one another. For example, measuring the position of a particle is associated in quantum mechanics with applying the position operator to the wave function. Measuring the momentum of a particle is understood in quantum mechanics as applying the partial-derivative-with-respect-to-position operator to the wave function (momentum, p, is classically the mass of the particle times its velocity, and velocity is defined as the derivative of position with respect to time). The two operators, position and momentum, *do not commute with each other*. That means that we cannot measure them both together, because if we measure one of them and then the other, our result would be different from what it would be if the order were reversed. The reason, in this example, for the noncommutativity of the two operators, position and momentum, can be seen by anyone who knows a little calculus: Derivative($X(\Psi)$)=Ψ +X(Derivative Ψ), which does not equal X(Derivative Ψ), which is an application of the two operators in the reverse order. The reason for the first expression above is the rule for taking the derivative of a product.

The fact that the two operators, X (position of the particle) and Derivative (momentum of the particle) do not commute has immense consequences in quantum mechanics. It tells us that we cannot measure both the position and the momentum of the same particle and expect to get good accuracy for both. If we know one of them to good precision (the one we measure *first*), then the other one will be known with poor precision. This fact is a mathematical consequence of the noncommutativity of the operators associated with these two kinds of measurements. This fact, that the position and the momentum of the same particle cannot both be localized with high precision is called the *uncertainty principle*, and it was also discovered by Werner Heisenberg. Heisenberg's uncertainty principle is his second important contribution to quantum theory after his formulation of matrix mechanics.

Heisenberg's uncertainty principle is fundamental to quantum mechanics and brings into quantum mechanics the ideas of probability theory on a very basic level. It states that uncertainty cannot be removed from quantum systems. The uncertainty principle is stated as:

$$\Delta p \Delta x \geq h$$

Here Δp is the difference in, or uncertainty about, momentum. And Δx is the difference in, or uncertainty about, location. The principle says that the product of the uncertainty in the position of a particle and the uncertainty in the momentum of the particle is greater than or equal to Planck's constant. The implications of this seemingly-simple formula are immense. If we know the position of the particle to very high precision, then we *cannot* know its momentum better than some given level of accuracy, *no matter how hard we try and no matter how good our instruments may be.* Conversely, if we know the momentum of a particle to good accuracy, then we cannot know the position well. The uncertainty in the system can never go away or be diminished below the level prescribed by Heisenberg's formula.

To demonstrate the uncertainty principle as applied to the position and momentum of a particle, we use *Heisenberg's Microscope.* In February of 1927, Bohr left Heisenberg to work alone in Copenhagen, and went skiing with his family in Norway. Being alone allowed Heisenberg's thoughts to wander freely, as he later described it, and he decided to make the uncertainty principle the central point of his interpretation of the nascent quantum theory. He remembered a discussion he'd had long before with a fellow student at

Göttingen, which gave him the idea of investigating the possibility of determining the position of a particle with the aid of a gamma-ray microscope. This notion solidified in his mind the principle he had already derived without this analogy. Heisenberg quickly wrote a letter to Wolfgang Pauli (another pioneer of the quantum theory) describing his thought-experiment about the use of a gamma-ray microscope to determine the position of a particle, and when he received Pauli's answer, he used the ideas in the letter to improve the paper he was writing. When Bohr returned from Norway, Heisenberg showed him the work, but Bohr was dissatisfied. Bohr wanted Heisenberg's argument to emanate from the duality between particles and waves. After a few weeks of arguments with Bohr, Heisenberg conceded that the uncertainty principle was tied-in with the other concepts of quantum mechanics, and the paper was ready for publication. What is Heisenberg's microscope? The figure below shows the microscope. A ray of light is shone on a particle and is reflected into the lens. As the ray of light is reflected by the particle into the microscope, it exerts on the particle it illuminates some pressure, which deflects the particle from its expected trajectory. If we want to lower the effect of the impact on the particle, so as not to disturb its momentum much, we must increase the wavelength. But when the wavelength reaches a certain amount, the light entering the microscope misses the position of the particle. So, one way or the other, there is a minimum to the level of accuracy that is possible to obtain for the *product* of position and momentum.

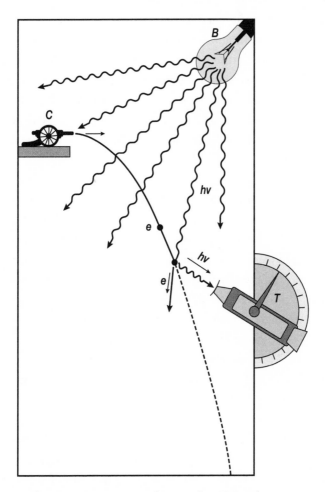

Another important contribution by Heisenberg to quantum mechanics was his discussion of the concept of *potentiality* within quantum systems. What separates quantum mechanics from classical mechanics is that in the quantum world a potential is always present, in addition to what actually happens. This is very important for the understanding of entanglement. The phenomenon of entanglement is a quantum phenomenon, and has no classical analog. It is the existence of potentialities that creates the entanglement. In particular, in a system of two entangled particles, the entanglement is evident in the potential occurrence of both AB (particle 1 in state A and particle 2 in state B) *and* CD (particle 1 in state C and particle 2 in state D). We will explore this further.

The 1930s brought great changes to the life of Heisenberg and to science. In 1932, Heisenberg was awarded the Nobel Prize for his work in physics. The next year, Hitler came to power and German science began its collapse with Jewish academics being fired by the Nazis. Heisenberg remained in Germany, watching his friends and colleagues leave for America and elsewhere. In an infamous SS paper, Heisenberg was reviled as a "white Jew," and "a Jew in spirit, inclination, and character," presumably for his apparent sympathy for his Jewish colleagues. But Heisenberg remained in Nazi Germany despite calls from his colleagues to leave. Just where his real sympathies were remains a mystery. There has been a suggestion that there were ties between Heisenberg's and Himmler's families.[10] It has been suggested that Heisenberg used that relationship to appeal directly to the leadership of the SS to end the diatribe against him. In 1937, the thirty-five year old Heisenberg, suffering from depression, met a twenty-two-year-old woman in a Leipzig bookshop. The two shared an interest in music, and performed together, he singing and she accompanying him on the piano. Within three months they were engaged, and shortly afterwards they were married.

In 1939, Heisenberg was called up for military service. By then he was the only leading physicist in Germany, and it was no surprise that for his military service the Nazis expected him to help them develop a nuclear bomb. In 1941, Heisenberg and his colleagues built a nuclear reactor, which they hid inside a cave under a church in a small village. Fortunately for humanity, Hitler's main project was Peenemunde, the Nazi effort to build missiles, which were directed at Britain, and the nuclear project was lower on the list of priorities. As it turned out, Heisenberg did not know how to make an atomic bomb, and the Manhattan Project in America was far ahead of the Nazi effort. After the war, Heisenberg remained a leading scientist in Germany, and he probably took with him to his grave the answers to the many questions humanity now has about the real role he played in the Nazi attempt to build a bomb.

9

Wheeler's Cat

"We will first understand how simple the universe is when we recognize how strange it is."

—John Archibald Wheeler

Many books on quantum mechanics tell the story Schrödinger used to illustrate a paradox based on the superposition of states. This story has come to be known as "Schrödinger's Cat." Schrödinger imagined a cat placed in a closed box along with an apparatus containing a speck of radioactive material. Part of the apparatus is a detector that controls a mechanism to break an attached vial of cyanide. When an atom of the radioactive element undergoes a disintegration that is registered by the detector, the vial is broken, killing the cat. Because the radioactive disintegration is a *quantum event*, the two states—the cat alive, and the cat dead—can be superimposed. Thus, before we open the box and make a measurement—i.e., before we actually find out whether the cat is dead or alive—the cat is both dead and alive at the same time. Besides the unpleasant implications, this example is not terribly instructive. Murray Gell-Mann says in his book *The Quark and the Jaguar* that Schrödinger's Cat is no better an example than that of opening the box containing a cat that has spent a long flight in the baggage hold of a plane. The owner of the arriving cat inevitably asks the terrifying question upon receiving the container at the baggage claim: Is my cat dead or alive? According to Gell-Mann the problem with the Schrödinger Cat example is that of *decoherence*. A cat is a large, macroscopic system, not an element of the microscopic quantum world. As such, the cat interacts with its environment very extensively: it breathes air, it absorbs and emits heat radiation, it eats and drinks. Therefore, it is impossible for the cat to behave in the very specific quantum way, in which it hangs on the balance "both dead and alive," as an electron in a superposition of more than one state.

I still like using a cat to make this point, but we don't have to have it dead, so our example will not be macabre. We will think of a cat being at two places at once, just as the electron. Think of an electron as a cat, Wheeler's Cat.

John Archibald Wheeler had a cat that lived with him and his family in Princeton. Einstein's house was only a short distance away, and the cat seemed to like Einstein's house. Wheeler would often see Einstein walking home, flanked by his two assistants, and sure enough, within a few minutes the phone would ring and Einstein would be on the line asking him whether he wanted him to bring back his cat. Imagine a cat that—instead of being dead and alive at the same time as in Schrödinger's example—is in the superposition of being both at Einstein's house and at Wheeler's house. When we take a measurement: Einstein or Wheeler looking for the cat, the cat is forced onto one of the two states, just as a particle or a photon.

The idea of a superposition of states is important in quantum mechanics. A particle can be in two states at the same time. Wheeler's cat, let us assume, can be in a superposition of the two states. The cat can be *both* at Wheeler's house *and* at Einstein's house. As Michael Horne likes to point out, in quantum mechanics we abandon the quotidian "either-or" logic in favor of the new "both-and" logic. And indeed, the concept is very foreign since we never encounter it in our daily lives. There are, perhaps, still some examples that can be made. I'm at the bank, and there are two lines in front of the teller windows. They're both equally long, and there's no one behind me. I want to be in the line that moves the fastest, but I don't know which one that will be. I stand between the two lines, or I keep jumping from one line to the other as one and the other becomes shorter. I am in "both lines at once." I am in a superposition of the two states: (I'm in line 1) and (I'm in line 2). Returning to Wheeler's cat, the cat is in a superposition of the following two states: (Cat at Wheeler's house) and (Cat at Einstein's house). Of course, in the original Schrödinger's Cat story, the cat is in a sadder superposition: (Cat is dead) and (Cat is alive).

John Archibald Wheeler was born in Jacksonville, Florida, in 1911. He received his doctorate in physics from Johns Hopkins University in 1933, and he also studied physics with Niels Bohr in Copenhagen. He took a professorship of physics at Princeton University, and his star student there was Richard Feynman (1918-1988). Feynman, who years later won a Nobel Prize and became one of the most famous American physicists, wrote his brilliant dissertation under Wheeler, leading to his doctorate from Princeton in 1942. The thesis, growing out of earlier work by Paul A. M. Dirac, introduced an important idea into quantum mechanics. It was an application of the classical principle of least action to the quantum world. What Feynman did was to create the sum-over-histories approach to quantum mechanics. This approach considers all possible paths a particle (or system) may take in going from one point to another. Each path has its own probability, and it is therefore possible to discover the most probable path the particle has taken. In Feynman's formulation, the wave-amplitudes attached to all possible paths are used to derive a total amplitude, and hence a probability distribution, for the outcomes at the common terminus of all possible paths.

Wheeler was very excited about Feynman's work, and took the manuscript of Feynman's thesis to Einstein. "Isn't this wonderful?" he asked. "Doesn't this make you believe in the quantum theory?" Einstein read the thesis, thought about it for a moment, and then said: "I still don't believe that God plays dice . . . but maybe I've earned the right to make my mistakes."[11]

Paul A.M. Dirac (1902-1984) was a British physicist who started his career as an electrical engineer. Since he had difficulties finding employment in his field, he applied for a fellowship at Cambridge University. Eventually he became one of the key figures in physics in the twentieth century and won the Nobel Prize. Dirac developed a theory that combined quantum mechanics with special relativity. His work thus allowed the equations of quantum mechanics to be corrected for relativistic effects for particles moving at close to the speed of light. As part of his research, Dirac predicted the existence of *anti-particles.* Dirac's paper on the theoretical possibility that anti-particles may exist was published in 1930, and a year later the American physicist Carl Anderson discovered the *positron,* the positively charged anti-electron, while analyzing cosmic rays. The electron and the positron annihilate each other when they meet, producing two photons.

In 1946, Wheeler proposed that the pair of photons produced when a positron and an electron annihilate each other could be used to test the theory of quantum electrodynamics. According to this theory, the two photons should have opposite polarizations: if one is vertically polarized, then the other one must be polarized horizontally. "Polarization" means the direction in space of either the electric or the magnetic fields of light.

In 1949, Chien-Shiung Wu (known as "Madame Wu," echoing the way physicists referred to Marie Curie) and Irving Shaknov, of Columbia University, carried out the experiment that Wheeler had suggested. Wu and Shaknov produced *positronium,* an artificial element made of an electron and a positron circling each other. This element lives for a fraction of a second, and then the electron and positron spiral toward each other, causing mutual annihilation that releases two photons. Wu and Shaknov used anthracene crystals to analyze the polarization direction of the resulting photons. Their result confirmed Wheeler's prediction: the two

photons were of opposite polarization. The 1949 Wu and
Shaknov experiment was the first one in history to produce
entangled photons, although this important fact was only
recognized eight years later, in 1957, by Bohm and
Aharonov.

Wheeler made important contributions to many areas of
physics in addition to quantum mechanics. These include the
theories of gravitation, relativity, and cosmology. He invented
the term *black hole* to describe the spacetime singularity that
results when a massive star dies. Together with Niels Bohr,
Wheeler discovered fission. In January 2001, at age 90,
Wheeler suffered a heart attack. The illness changed his view
of life, and he decided that he wanted to spend his remaining time working on the most important problems in physics:
the problems of the quantum.

According to Wheeler, the problem of the quantum is the
problem of being, of existence. He vividly recalls the story,
related by H. Casimir, a fellow student of Bohr, of the debate
on the quantum between Bohr and Heisenberg. The two
were invited by the philosopher Høffding, a mutual friend, to
his home to discuss the Young double-slit experiment and its
implications about the quantum. Where did the particle go?
Did it pass through one slit or the other? As the discussion
progressed, Bohr mulled the issue and muttered: "To be . . .
to be . . . what does it *mean* to be?"

John Wheeler himself later took the Young double-slit
experiment to a new level. He showed in a cogent and elegant
way that within a variant of this experiment, with the mere
act of measurement, an experimenter can change *history*. By

deciding whether we want to measure something one way, or another, the experimenter, a human being, can determine what "shall have happened in the past." The following description of Wheeler's experimental setup is adapted from his paper, "Law without Law."[12]

Wheeler described in the article a modern variant of Young's double-slit experimental setup. The figure below shows the usual double-slit arrangement.

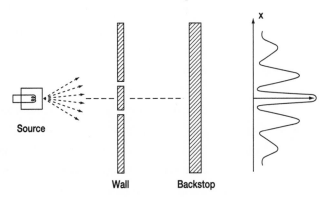

Light rays hit the screen with the slits and produce two sets of waves, as would happen with water waves as they leave the two slits. Where they meet, the light waves interact with each other. They interfere constructively to result in a higher amplitude wave, and destructively, canceling each other or producing waves of lower amplitude. The modern setup uses mirrors instead of slits, and it uses laser light, which can be controlled much more precisely than regular light. In more advanced experimental arrangements, fiber optics are used as the medium of choice for experiments.

The simplest of the setups for the analog of the double-slit experiment is shown below. We have a diamond design, in which light from a source is aimed at a half-silvered mirror that allows half the light to go through and half to be reflected. Such a mirror is called a *beam splitter* since it splits an arriving light beam into two beams: the reflected beam and the transmitted beam. The beams are then each reflected by a mirror and are allowed to cross and then be detected. By noting which detector clicks to register the arrival of a photon, the experimenter is able to tell by which route the photon has traveled: Was the photon transmitted by the beam splitter, or was it reflected through it? Alternatively, the experimenter can place another beam splitter (half-silvered mirror) right at the point of crossing of the two beams. Such a placement of the mirror will cause the beams to interfere with each other, just as in the double-slit experiment. Here, only one detector will click (where the beams interfere constructively) and the other will not (since there the beams interfere destructively). When this happens in an experiment with very weak light that sends only one photon at a time, we find that the photon travels *both* paths—it is both reflected and transmitted at the first beam splitter (or else there would be no interference: both detectors would click, which does not happen).

Wheeler says that Einstein, who used a similar idea in a thought experiment, argued that "it is unreasonable for a single photon to travel simultaneously two routes. Remove the half-silvered mirror, and one finds that one counter goes off, or the other. Thus the photon has traveled only one route. It travels only one route, but it travels both routes; it travels both routes, but it travels only one route. What nonsense! How obvious it is that quantum theory is inconsistent! Bohr emphasized that there is no inconsistency. We are dealing with two different experiments. The one with the half-silvered mirror removed tells which route. The one with the half-silvered mirror in place provides evidence that the photon traveled both routes. But it is impossible to do both experiments at once."[13]

Wheeler asked the question: Can the experimenter determine by which route the photon travels? If the experimenter leaves out the second beam splitter, then the detectors indicate by which route the photon traveled. If the second beam splitter is in place, we know from the fact that one detector clicks and not the other that the photon has traveled both paths. Before the decision is made whether to insert the beam splitter, one can only describe the photon in the interferometer as being in a state with several potentialities (since potentialities can coexist). The choice of inserting or not inserting the beam splitter determines which potentiality is actualized. The two setups are shown below.

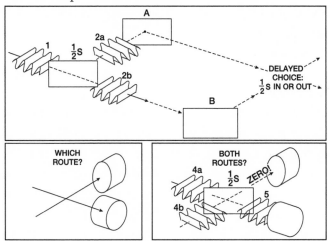

The amazing thing, according to Wheeler, is that by *delayed choice*, the experimenter can change history. The experimenter can determine whether or not to place the second beam splitter *after the photon has traveled most of the way to its destination.* Modern science allows us to randomly choose the action (place the beam splitter or not place it) so quickly—within a very, very small fraction of a second—so that the photon will have already done its travel. When we do so, we are determining, *after the fact*, which route the photon shall have traveled. Shall it have traveled one route, or shall it have traveled both routes?

Wheeler then took his outlandish idea to the cosmic scale.[14] He asked: "How did the universe come into being? Is that some strange, far-off process, beyond hope of analysis? Or is the mechanism that came into play one which all the time shows itself?" Wheeler thus tied the big bang and the creation of the universe to a quantum event, and he did so years before the cosmologists of the 1980s and 1990s came up with the idea that the galaxies were formed because of quantum fluctuations in the primordial soup of the big bang. Wheeler's answer to creation, history, and the birth of the universe is that we should look at the delayed-choice experiment. Such an experiment "reaches back into the past in apparent opposition to the normal order of time." He gives the example of a quasar, known as 0957+561A,B, which scientists had once thought to be two objects but now believe is one quasar. The light from this quasar is split around an intervening galaxy between us and the quasar. This intervening galaxy acts as a "gravitational lens," about which the light of the quasar is split. The galaxy takes two light rays,

spread apart by fifty thousand light years on their way from
the quasar to Earth, and brings them back together on Earth.
We can perform a delayed-choice split-beam experiment with
the quasar acting as the half-silvered mirror and the intervening galaxy as the two full mirrors in the experimental
setup used in the laboratory. Thus we have a quantum experiment on a cosmic scale. Instead of a few meters' distance, as
in the lab, here we have an experiment with distances of billions of light years. But the principle is the same.

Wheeler says: "We get up in the morning and spend the day
in meditation whether to observe by 'which route' or to
observe interference between 'both routes.' When night comes
and the telescope is at last usable, we leave the half-silvered
mirror out or put it in, according to our choice. The mono-
chromatizing filter placed over the telescope makes the counting rate low. We may have to wait an hour for the first photon.
When it triggers a counter, we discover 'by which route' it
came with the one arrangement; or by the other, what the relative phase is of the waves associated with the passage of the
photon from source to receptor 'by both routes'—perhaps
50,000 light years apart as they pass the lensing galaxy g-1. But
the photon has already *passed* that galaxy billions of years
before we made our decision. This is the sense in which, in a
loose way of speaking, we decide what the photon *shall have
done* after it has *already* done it. In actuality it is wrong to talk
of the 'route' of the photon. For a proper way of speaking we
recall once more that it makes no sense to talk of the phenomenon until it has been brought to a close by an irreversible
act of amplification: 'No elementary phenomenon is a phenomenon until it is a registered (observed) phenomenon.'"

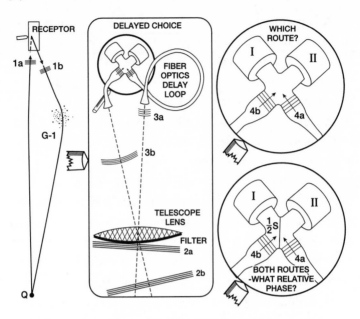

Reproduced from J.A. Wheeler, *Law without Law,*
Wheeler and Zureck, eds., 1983.

10

The Hungarian Mathematician

"I do know, though, that when in Princeton Bohr would
often discuss measurement theory with Johnny von Neumann who pioneered the field. As I see it, these considerations have led to contributions, important ones, to
mathematics rather than to physics."

—Abraham Pais

Jancsi ("Johnny") Neumann was born in Budapest on
December 28, 1903, to a wealthy family of bankers.
Between 1870 and 1910, Budapest saw an unprecedented economic boom, and many talented people migrated
there from the Hungarian countryside and from other nations
to pursue the opportunities that this thriving European capital had to offer. By 1900 Budapest boasted 600 coffeehouses,
numerous playhouses, a renowned symphony and opera company, and an educational system that was the envy of Europe.
Ambitious, hard-working people thronged to Budapest, where
they had a chance of achieving success in the growing economic life of the city. Among the newcomers were many Jews
who flocked from all over Europe to a city known for its reli-
gious tolerance and enlightened population.[15]

Johnny's parents, Max and Margaret Neumann, came to

Budapest from the town of Pecs, on the Yugoslav border, as many Jews did during that time, the late 1800s. Max worked hard but was handsomely rewarded and within a few years became a powerful executive at a successful Hungarian bank that prospered from lending money to small business owners as well as agricultural corporations. Max did so well, in fact, that within a few years he was able to buy his family an 18-room apartment in a building in which several other wealthy Jewish families resided. One of them was the family of his brother-in-law. The children of the two families roamed the floors of the building together, running in and out of both palatial apartments.

In addition to financial success, Max Neumann achieved a degree of influence over Hungarian politics. As a major figure in Hungarian society and a successful financial advisor to the Hungarian Government, Max Neumann was rewarded with a hereditary nobility title in 1913. This was the Hungarian equivalent of being knighted by the Queen of England. In addition to the great honor—rare for Jews—Max could now add the prefix *von* to his name. He became Max von Neumann, a member of the Hungarian nobility. His sons, John, the firstborn, and his two brothers, Michael and Nicholas, enjoyed the same privilege. At age 10, the young Jancsi Neumann thus became John von Neumann, and throughout his life he cherished his noble European status. The family even drew its own coat of arms, including a rabbit, a cat, and a rooster. Max thought that Johnny was like a rooster because he sometimes used to crow; Michael looked like a cat; and Nicholas, the youngest, was the rabbit. The von Neumanns exhibited their coat of arms at their impres-

sive apartment in the city and also on the gate of the sump-
tuous country estate they later bought, where they spent their
summers. The family not only became members of the Hungarian aristocracy, they ranked among its staunchest defend-
ers. After Bela Kun established Communist rule in 1919,
Max von Neumann went to Vienna and summoned Admiral
Horthy to attack Kun's forces and retake Hungary, freeing it
from Communism for the first time (this happened a second
time after the collapse of the Soviet Union).

In the fateful year 1913, when the family received its nobility and war was declared across Europe, Johnny began to
exhibit the startling intellectual capacity that in time separated
him from the rest of his family, and everyone else around him.
The discovery was made, innocently enough, when his father
asked the ten-year-old to multiply two numbers and the boy
completed the task with amazing speed. Max then gave Johnny
two huge numbers to multiply, and the child completed the
calculation in seconds. This stunned the father and he began
to realize that this was no ordinary child. Johnny was gifted far
beyond what anyone had imagined.

Only later came the revelations that at school Johnny knew
more about the material he was taught than did his teachers.
In conversations around the family dinner table he was far
ahead of all the other family members in his understanding
of the issues and ideas discussed.

When his parents understood that their firstborn child was
so prodigiously gifted, they did not waste the opportunity to
prepare him for greatness. They hired private tutors to teach
him advanced mathematics and science. And the father led
intellectual discussions around the dinner table in which

every family member was expected to contribute to the discussion. This allowed the young genius to further refine his skills.

At age eleven, Johnny was sent to the gymnasium, the European institution similar to a high school, which normally accepted students several years older. At the gymnasium Johnny studied mathematics, Greek, Latin, and other subjects. He excelled in all of them. Laslo Ratz, an instructor of mathematics at the gymnasium, quickly realized that he had a genius in his class. He went to Max von Neumann and suggested that the family provide their son with even more training in mathematics. It was arranged that Ratz would take Johnny out of the regular math class three times a week and teach him privately. But soon Ratz realized that Johnny knew more than he did. Ratz took Johnny to the University of Budapest. Here the boy—clearly the youngest person ever to attend the university—enrolled in classes in advanced mathematics.

A year after he started taking classes at the university, a fellow student (years older) asked Johnny if he had heard of a particular theorem in number theory. Johnny knew the theorem—it was an unproven result that many mathematicians had worked on. His friend (who years later won a Nobel Prize) asked him if he could prove it. Johnny worked on the theorem for several hours, and proved it. Within a year he enrolled at the renowned technical university in Zurich, the ETH (Einstein's alma mater), and a short time later at the University of Berlin. At all three institutions he stunned renowned mathematicians, among them the famous David Hilbert (1862-1943), with his keen understanding of mathematics and his incredible ability to compute and analyze problems with unparalleled speed.

When solving a mathematical problem, von Neumann would face a wall, his face would lose all expression, and he would mutter to himself for several minutes. Completely immersed in the problem, he would not hear or see anything that was happening around him. Then suddenly his face would assume its normal expression, he would turn back from the wall, and quietly state the answer to the problem.

Johnny von Neumann was not the only outstanding intellect that Budapest produced during those years. Six Nobel Prize winners were born in Budapest between 1875 and 1905 (five of them Jewish). Four other leaders of modern science and mathematics were also born in Budapest during this period. All of them had attended the superb schools of Hungary, the gymnasia, and were nurtured at home. Half a century later, Nobel laureate Eugene Wigner, who was one of these ten geniuses, was asked what he thought was the reason for the phenomenon. Wigner replied that he didn't understand the question. "Hungary has produced only one genius," he said. "His name is John von Neumann."

Most of the Hungarian prodigies emigrated to the United States, where their influence on the development of modern science was immense. When they arrived in America, their special gifts stunned the scientific community, and some began to half-seriously speculate that the foreign scientists were not Hungarian, but aliens from outer space bent on dominating American science. Theodore von Karman was the first of the ten to come to America. He was followed by Edward Teller and the others, including John von Neumann, in the 1930s. When Teller arrived, he was confronted with the story about the extraterrestrial origin of these geniuses. Teller assumed a worried expression. Then he said: "von Karman must have been talking."

But before immigrating to the United States, Johnny von Neumann—arguably the greatest genius of them all—received further superb training in mathematics and science that helped transform him into one of the greatest mathematicians of his age. This training took place at the universities of Zurich, Göttingen, and Berlin.

In 1926, von Neumann came to Göttingen and heard a lecture by Werner Heisenberg on matrix mechanics and the difference between his approach to quantum mechanics and that of Schrödinger (roughly the same talk the author heard in Berkeley 46 years later). In the audience was also David Hilbert, the greatest mathematician of the time. According to Norman Macrae (*John von Neumann*, AMS, 1999), Hilbert didn't understand the quantum theory as presented by Heisenberg and asked his assistant to explain it. Von Neumann saw this and decided to explain quantum theory to the old mathematician in terms that he could understand, that is, in mathematical language. In doing so, von Neumann used the ideas of *Hilbert Space*, much to the delight of Hilbert.

To this day, physicists use Hilbert space to explain and analyze the world of the very small. A Hilbert space is a linear vector space with a norm (a measure of distance) and the property of completeness.

Von Neumann expanded the paper he wrote for Hilbert in 1926 into a book called *The Mathematical Foundations of Quantum Mechanics*, published in 1932. Von Neumann demonstrated that the geometry of vectors over the complex plane has the same formal properties of the states of a quantum mechanical system. He also derived a theorem, using some assumptions about the physical world, which proved that there are no "hidden variables," whose inclusion could reduce the uncertainty in quantum systems. While posterity would agree with his conclusion, John Bell successfully challenged von Neumann's assumptions in his daring papers of the 1960s. Still, von Neumann was one of the founders of the mathematical foundations of the quantum theory, and his work is important in establishing mathematical models for the inexplicable physical phenomena of the quantum world. Key among these concepts is the idea of a Hilbert space.

A Hilbert space, denoted by H, is a complete linear vector space (where *complete* means that sequences of elements in this space converge to elements of the space). As applied in physics, the space is defined over the complex numbers, which is needed in order to endow the space with the necessary richness for making it the proper model in different situations. Complex numbers are numbers that may contain the element i, the square root of negative one. The Hilbert space H allows the physicist to manipulate vectors, that is, mathematical entities that have both a magnitude and a direction: little arrows in Hilbert space. These arrows can be added to or subtracted from each other, as well as multiplied by numbers. These arrows are the mathematical essentials of the physical theory since they represent states of quantum systems.

Von Neumann came to the Institute for Advanced Studies at Princeton in the early 1930s. He and Einstein never saw eye to eye. Their differences were mostly political—von Neumann found Einstein naïve, believing, himself, that all left-leaning governments were weak, and dogmatically supporting conservative policies. He was involved with the Manhattan Project and, unlike most other scientists who contributed to the making of the bomb, never appeared to have battled any moral dilemmas as a result of this work.

No one questions the fact that von Neumann made great contributions to the quantum theory. His book on the subject has become an indispensable tool for practitioners and an important treatise on the mathematical foundations of quantum mechanics.

Eugene Wigner, who later won the Nobel Prize for his work in physics, came to Princeton after John von Neumann was already established there. Some have said that Wigner was hired by Princeton from Hungary so that "Johnny" wouldn't be lonely and would have someone who could speak Hungarian with him. When von Neumann's seminal book appeared in English, Wigner told Abner Shimony: "I have learned much about quantum theory from Johnny, but the material in his Chapter Six [on measurement] Johnny learned all from me." Von Neumann's book contained an argument that was important in subsequent discussions of the interpretations of quantum mechanics, namely a proof that the quantum theory could not be "completed" by a hidden-variables theory in which every observable has a definite value. His proof of this proposition was mathematically correct, but was based on a premise that is dubious from a physical point of view. This flaw in von Neumann's book was exposed decades later by John Bell.

11

Enter Einstein

"The elementary processes make the establishment of a truly quantum-based theory of radiation appear almost inevitable."

—Albert Einstein

Albert Einstein was born in Ulm, in southern Germany, in 1879 to a middle-class Jewish family. His father and uncle owned an electrochemical business, which kept failing. As a result, the family moved to Munich, then to a couple of places in northern Italy, and finally back to Germany. Einstein was educated in Switzerland, and his first job was famously that of technical expert at the Swiss Patent Office in Bern. There, in the year 1905, Einstein published *three* papers that changed the world. These papers were his expositions of the three theories he developed while working alone at the patent office: the special theory of relativity; a theory of Brownian motion and a new formulation of statistical thermodynamics; and a theory of the photoelectric effect.

Einstein's life and his development of the theories of relativity have been discussed in detail.[16] But Einstein exerted a very important influence on the quantum theory from its inception. Soon after he read Planck's paper about the quantum in 1900, Einstein began to ponder the nature of light in view of the new theory. He proposed the hypothesis that light is a stream of particles, or *quanta*.

Einstein studied the effect of the interaction of light with matter. When light rays strike a metal, electrons are emitted. These electrons can be detected and their energies measured. This was proved using a number of experiments by the American physicist Robert Millikan (1868-1953). The analysis of the photoelectric effect in various metals and using light of different frequencies revealed the following phenomena: When light of low frequency, up to a threshold frequency v_0, shines on a metallic surface, no photoelectrons are emitted. For a frequency above the threshold, photoelectrons are emitted and as the intensity of the light of this frequency is varied, the number of photoelectrons varies but their energy remains the same. The energy of the photoelectrons increases only if the frequency is increased. The threshold frequency, v_0, depends on the kind of metal used.

The classical theory of light does not explain the above phenomena. Why shouldn't the intensity of light increase the energy of the photoelectrons? And why would the frequency affect the energy of these electrons? Why are no photoelectrons released when the frequency is below some given level? What Einstein did in his research culminating with the paper of 1905 was to assume that light consisted of particles—later called photons—and to apply Planck's quantum idea to these photons.

Einstein viewed the photons as discrete little packages of

energy flying through space. Their energy was determined by Einstein's formula: $E=h\nu$. (Where h is Planck's constant and ν is the frequency of the light.)

The connection between this formula and Planck's earlier equation is simple. Recall that Planck had said that the only possible energy levels for a light-emitting system (i.e., an oscillating charge) are:

$E=0$, $h\nu$, $2h\nu$, $3h\nu$ $4h\nu$ . . . , or in general, $nh\nu$, where n is a positive integer.

Clearly, the smallest amount of energy that can be emitted by the system is the difference between two adjacent Planck values, which is: $h\nu$, hence Einstein's formula for the energy of the smallest possible amount of light.

We see from Einstein's formula that the intensity of the light does not increase the energy of its photons, but only increases the *number* of photons emitted, the energy of each photon being determined by the frequency of the light (multiplied by Planck's constant). In order to disengage an electron from the lattice of atoms in the metal, some minimum energy is required, denoted by W (which stands for "work"—the work necessary to dislodge one electron). Thus when the frequency reaches some minimum level, the energy imparted to the electron passes the threshold W, and the electron is released. Einstein's law for explaining the photoelectric effect is given by the formula:

$$K=h\nu-W$$

Where K is the kinetic energy of the released electron. This energy is equal to Einstein's energy ($E=h\nu$) minus the minimum level needed to dislodge the electron, *W*. The formula explained the photoelectric effect perfectly. This elegant theory of the interaction of light with matter, a quantum theory of a known and previously-misunderstood effect, won Einstein the Nobel Prize in 1921. He was notified of the Prize while on a visit to Japan. Curiously, Einstein never received a Nobel Prize for his special theory of relativity, nor for his general theory of relativity, two theories that revolutionized modern science.

So Einstein was there when the quantum theory was born, and was one of the "fathers" of the new theory. He felt he understood nature very well, as evidenced by the fact that he could propose such revolutionary theories—his special theory of relativity of 1905 and the general theory of relativity of 1916—that explained ever more phenomena in the realm of the large and fast. But even though he was an incomparable master of the physics of the macro-world, and contributed much to the quantum theory of the very small, Einstein's philosophy clashed with the advancing interpretation of quantum mechanics. Einstein could not give up his belief that God does not play dice, meaning that chance has no place within the laws of nature. He believed that quantum mechanics was correct to attribute probabilities to possible outcomes of an experiment, but he thought that the need to resort to probabilities was due to our ignorance of a deeper level of the theory, which is describable by deterministic (i.e., devoid of a probability-structure) physics. That is the meaning of his oft-quoted statement about God and dice.

Quantum theory was—and still is—based on probabilities rather than exact predictions. As the Heisenberg uncertainty principle specifies, it is impossible to know both the momentum of a particle and its location—if one is known with some precision, the other, of necessity, can only be known with uncertainty. But the randomness, the variation, the fuzziness, the uncertainty in the new physical theory goes beyond its manifestation in the uncertainty principle. Recall that particles and photons are both wave and particle and that each has its wave function. What is this "wave function"? It is something that leads *directly* to probabilities, since the square of the amplitude of the wave we associate with any particle *is*, in fact, a probability distribution (the rule that assigns probabilities of various outcomes) for a particle's position. To obtain the probability distribution of the outcome of a measurement of other observable quantities (such as momentum), the physicist must perform a calculation that uses both the wave function and the operator representing the observable quantity of interest.

Quantum theory is probabilistic on a very basic, very fundamental level. There is no escape from the probabilities regardless of what we do. There is a minimum level of uncertainty about outcomes, which can never be diminished, according to the theory, no matter how hard we try. The quantum theory is thus very different from other theories that use probabilities. In economics, for example, there is no theory that states unequivocally that we cannot know some variable to a level of precision we desire. Here, the probabilities represent our lack of knowledge, not a fundamental property of nature. Einstein was a great critic of the quantum theory because he did not like to think that nature works probabilistically. God decrees, He does not play dice. Thus,

Einstein believed that there was something missing from the quantum theory, some variables, perhaps, such that if we could find the values of these variables, the uncertainty—the randomness, the "dice"—would be gone. With the augmentation of these variables, the theory would be complete and would thus be like Newton's theory, in which variables and quantities may be known and predicted with great precision.

In addition to his dislike for randomness and probability within a theory of nature, Einstein had other notions—ones that were "intuitive" to him, and would be so to most people. These were notions of realism and of locality. To Einstein, a facet of reality is something real, which a good theory of nature should include. If something happens somewhere, and we can predict it will happen without disturbing the system, then what happens is an element of reality. If a particle is located at a given spot, and we can predict that it will be there without disturbing it, then that is an element of reality. If a particle spins in a certain direction, and we can predict that it will spin in this direction without disturbing it, then this is an element of reality. Locality is the intuitive notion that something that happens in one place should not affect whatever happens at a far away location, unless, of course, a signal is sent to the other location (at the speed of light or less, as dictated by the special theory of relativity) and can make a difference there.

Throughout his life, Einstein held fast to these three principles he believed should be part of a good description of nature:

1. The fundamental level of nature should be described in principle by a deterministic theory, even though gaps in human knowledge about initial and boundary conditions may force human beings to resort to probability in making predictions about the outcomes of observations.

2. The theory should include all elements of reality.

3. The theory should be local: what happens here depends on elements of reality located here, and whatever happens there depends on elements of reality located there.

Einstein and his collaborators found that these notions, which seemed very natural to them, implied the incompleteness of the quantum theory—a theory that Einstein himself had helped bring about. As we will see, the above principles were eventually shown to be *incompatible* with quantum theory, but this would only happen in the 1960s. And mounting experimental evidence collected since the 1970s would further imply that quantum theory was correct.

In the spring of 1910, the Belgian industrialist Ernest Solvay came up with the idea of organizing a scientific conference. The route to this idea was somewhat circuitous and bizarre. Solvay had developed a method for manufacturing soda and as a result became very wealthy. This gave him a high level of confidence in his own abilities, and, since he was interested in science, he began to dabble in physics. Solvay invented a theory of gravitation and matter, which had little to do with reality or with science. But since he was so wealthy, people listened to him, even if they could tell that his theories were nonsensical. The German scientist Walther Nernst told Solvay that he could get an audience for his theories if he would organize a conference for the greatest physicists of the day, and present to them his theories. Solvay fell for the idea, and thus the Solvay Conferences were born.

The first *Conseil Solvay* took place at the Metropole Hotel in Brussels in late October 1911. Invitations were sent to the best-known physicists, including Einstein, Planck, Madame Curie, Lorentz, and others, and all the invitees accepted and attended what became a historic meeting. The conferences continued over the next two decades. Future meetings were the battlegrounds for the great controversy of the quantum theory. Here in Brussels, at the later conferences, Bohr and Einstein argued over the philosophical and physical implications of quantum mechanics.

Einstein had admired Bohr since the publication of Bohr's first paper on the quantum theory of atoms in 1913. In April 1920, Bohr came to Berlin to deliver a series of lectures. Einstein held a position in that city with the Prussian Academy of Science. The two men met, and Bohr spent some time with the Einsteins at their home. He had brought them gifts: good Danish butter and other foodstuffs. Einstein and Bohr enjoyed engrossing conversations on radiation and atomic theory. After Bohr left, Einstein wrote him: "Seldom in my life has a person given me such pleasure by his mere presence as you have. I am now studying your great publications and—unless I happen to get stuck somewhere—have the pleasure of seeing before me your cheerful boyish face, smiling and explaining."[17]

Over the years, their relationship matured into an amicable competition for the truth about nature. Bohr, the orthodox interpreter of the quantum theory, was defending its curious facets, while Einstein, the realist, was ever pushing for a more "natural" theory—one which, alas, neither he nor anyone else was able to produce.

The debate between Einstein and Bohr on the interpretation of quantum mechanics began in earnest during the fifth Solvay Conference in October 1927. All the founders of the quantum theory were there: Planck, Einstein, Bohr, de Broglie, Heisenberg, Schrödinger, Dirac. During the meetings, "Einstein said hardly anything beyond presenting a very simple objection to the probability interpretation . . . .Then he fell back into silence."[18] But in the dining room of the hotel, Einstein was very lively. According to a firsthand account by Otto Stern, "Einstein came down to breakfast and expressed his misgivings about the new quantum theory. Every time, he had invented some beautiful [thought] experiment from which one saw that it did not work. Pauli and Heisenberg, who were there, did not react well to these matters, 'ach was, das stimmt schon, das stimmt schon' ('ah well, it will be all right, it will be all right'). But Bohr, on the other hand, reflected on it with care, and in the evening, at dinner, we were all together and he cleared up the matter in detail."[19]

Heisenberg, an important participant in the 1927 conference, also described the debate: "The discussions were soon focused to a duel between Einstein and Bohr on the question as to what extent atomic theory in its present form could be considered to be the final solution of the difficulties which had been discussed for several decades. We generally met already at breakfast in the hotel, and Einstein began to describe an ideal experiment which he thought revealed the inner contradictions of the Copenhagen interpretation."[20]

Bohr would work all day to find an answer to Einstein, and by late afternoon he would show his argument to his fellow quantum theorists. At dinner, he would show Einstein his answer to Einstein's objection of the morning. Although Einstein would find no good objection to the argument, in his heart he remained unconvinced. According to Heisenberg, Einstein's good friend Paul Ehrenfest (1880-1933) told him: "I am ashamed of you. You put yourself in the same position as your opponents in their futile attempts to refute your relativity theory."

The arguments for and against the quantum theory intensified during the next Solvay Conference, which took place in 1930. The topic of the meeting was magnetism, but that did not prevent the participants from continuing their heated debate of the quantum theory outside the regular sessions, in corridors, and at the breakfast and dinner tables at the hotel. Once, at breakfast, Einstein told Bohr that he had found a counterexample to the uncertainty principle for energy and time. Einstein envisioned an ingenious, complex device: a box with an opening in one of its walls, where a door is placed, controlled by a clock inside the box. The box is filled with radiation and weighed. The door is opened for a split second, allowing one photon to escape. The box is again weighed. From the weight difference, one can deduce the energy of the photon by using Einstein's formula, $E=mc^2$. Thus, argued Einstein, in principle one can determine to any level of accuracy both the photon's energy and its time of passage, refuting the uncertainty principle (which says, in this context, that you cannot know both the time of passage and the energy to high precision). Einstein's device is shown below.

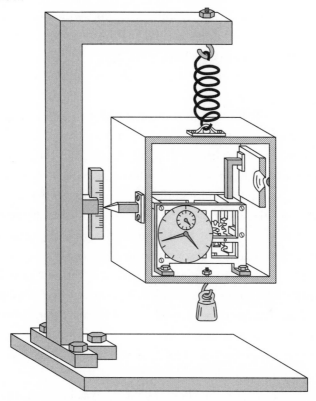

As reported by Pais (1991), participants at the conference found Bohr in shock. He did not see a solution to Einstein's challenge to quantum theory. During that entire evening he was extremely unhappy, going from one person to the next, trying to persuade them that Einstein's conclusion could not be true: but how? If Einstein was right, Bohr said, it would be the end of physics. Hard as he tried, however, he could not refute Einstein's clever argument. Leon Rosenfeld (1904-1974), a physicist present at the meetings, said: "I shall never forget the sight of the two antagonists leaving the club: Einstein a tall majestic figure, walking quietly, with a somewhat ironical smile, and Bohr trotting near him, very excited . . . . The next morning came Bohr's triumph."[21]

There is a picture that captures this description very well (See the first photograph, page 131.)

Bohr finally found a flaw in Einstein's argument. Einstein had failed to take account of the fact that weighing the box amounts to observing its displacement within the gravitational field. The imprecision in the displacement of the box generates an uncertainty in the determination of the mass—and hence the energy—of the photon. And when the box is displaced, so is the clock inside it. It now ticks in a gravitational field that is slightly different from the one it was in initially. The rate of ticking of the clock in the new position is different from its rate before it was moved by the weighing process. Thus there is an uncertainty in the determination of time. Bohr was able to prove that the uncertainty relationship of energy and time was exactly as stated by the uncertainty principle.

Bohr's answer to Einstein's challenge was brilliant, and it used Einstein's own general theory of relativity in parrying the attack. The fact that clocks tick at different rates depending on the gravitational field is an important facet of general relativity. Here Bohr used a clever argument in applying relativity theory to establish the quantum-mechanical uncertainty principle.

But the controversy raged on. Einstein, the wily fox of physics, kept coming up with increasingly more clever arguments in an effort to combat a theory whose very foundation he found upsetting. And since, as one of its founders, he knew quantum theory better than anyone, he knew how to deal his blows. Whenever Einstein would strike, Bohr would get upset and worry and frantically search for an answer until one was found. He would often repeat a word to himself while lost in thought. Fellow physicists reported him standing in a room, muttering: "Einstein . . . Einstein . . . ," walking over to the window, looking out, lost in thought, and continuing: "Einstein . . . Einstein . . . ."

Einstein attended the 1933 Solvay Conference, and heard Bohr give a talk about the quantum theory. He followed the argument attentively, but did not comment on it. When the discussion began, he led it in the direction of the meaning of quantum mechanics. As Rosenfeld described it, Einstein "still felt the same uneasiness as before ('unbehagen' was the word he used) when confronted with the strange consequences of the theory."[22] It was during this occasion that he first brought out what would later be seen as his most formidable weapon against the quantum theory. "What would you say of the following situation?" he asked Rosenfeld. "Suppose two particles are set in motion towards each other with the same, very large, momentum, and that they interact with each other for a very short time when they pass at known positions. Consider now an observer who gets hold of one of the particles, far away from the region of interaction, and measures its momentum; then, from the conditions of the experiment, he will obviously be able to deduce the momentum of the other particle. If, however, he chooses to measure the position of the first particle, he will be able to tell where the other particle is. This is a perfectly correct and straightforward deduction from the principles of quantum mechanics; but is it not very paradoxical? How can the final state of the second particle be influenced by a measurement performed on the first, after all physical interaction has ceased between them?"

Here it was, two years before it was unleashed on the world of science with all its might—Einstein's tremendously potent idea about quantum theory, in which he used the theory's apparent contradictions to invalidate itself. Rosenfeld, with whom Einstein shared his thought while listening to Bohr's presentation, did not think that Einstein meant more by this thought experiment than an illustration of an unfamiliar feature of quantum mechanics. But the spark of the idea Einstein first formulated during Bohr's presentation would continue to grow and would take its final form two years later.

When Hitler came to power, Albert Einstein left Germany. Already in 1930, Einstein had spent considerable portions of his time abroad: he was at Caltech in California, and later at Oxford University. In 1933, Einstein accepted a position at the newly established Institute for Advanced Study at Princeton. He had planned to spend some of his time there and some of it in Berlin, but following Hitler's victory, he quit all his appointments in Germany and vowed never to return. He spent some time in Belgium and England, and finally arrived at Princeton in October 1933.

Einstein settled in his new position at the Institute for Advanced Study. He was given an assistant, a twenty-four year old American physicist named Nathan Rosen (1910-1995). And he was reacquainted with a physicist at the Institute whom he had known at Caltech three years earlier, Boris Podolsky. Einstein may have moved across the Atlantic, thousands of miles away from the Europe in which the quantum theory was born and developed, but the outlandish theory with its incomprehensible logic and assumptions remained on his mind.

Einstein had usually worked alone, and his papers were rarely coauthored. But in 1934, he enlisted the help of Podolsky and Rosen in writing one last polemic against the quantum theory.[23] Einstein later explained how the now-famous Einstein, Podolsky, and Rosen (EPR) paper was written in a letter to Erwin Schrödinger the following year: "For linguistic reasons, the paper was written by Podolsky, after prolonged discussions. But what I really wanted to say hasn't come out so well; instead, the main thing is, as it were, buried under learning." Despite Einstein's impression to the contrary, the message of the EPR article, in which he and his colleagues used the concept of entanglement to question the completeness of the quantum theory, was heard loud and clear around the world. In Zurich, Wolfgang Pauli (1900-1958), one of the founders of the quantum theory and the discoverer of the "exclusion principle" for atomic electrons, was furious. He wrote a long letter to Heisenberg, in which he said: "Einstein has once again expressed himself publicly on quantum mechanics, indeed in the 15 May issue of *Physical Review* (together with Podolsky and Rosen—no good company, by the way). As is well known, every time that happens it is a catastrophe." Pauli was upset that the EPR paper was published in an American journal, and he was worried that American public opinion would turn against the quantum theory. Pauli suggested that Werner Heisenberg, whose uncertainty principle bore the brunt of the EPR paper, issue a quick rejoinder.

But in Copenhagen, the response was the most pronounced. Niels Bohr seemed as if hit by lightning. He was in shock, confused, and he was angry. He withdrew and went home. According to Pais, Rosenfeld was visiting Copenhagen at that time, and said that the next morning, Bohr appeared at his office all smiles. He turned to Rosenfeld and said: "Podolsky, Opodolsky, Iopodolsky, Siopodolsky, Asiopodolsky, Basiopodolsky . . . ." To the bewildered physicist he explained that he was adapting a line from the Holberg play *Ulysses von Ithaca* (Act I, Scene 15), in which a servant suddenly starts to talk gibberish.[24]

Rosenfeld recalled that Bohr abandoned all the projects he was working on when the EPR paper came out. He felt the misunderstanding had to be cleared up as quickly as possible. Bohr suggested that he and his helpers use the same example that Einstein used to demonstrate the "right" way to think about it. Bohr, excited, began to dictate to Rosenfeld the response to Einstein. But soon, he hesitated: "No, this won't do . . . we have to do this all over again . . . we must make it quite clear . . . ." According to Rosenfeld, this went on for quite a while. Every once in a while, Bohr would turn to Rosenfeld: "What *can* they mean? Do *you* understand it?" He would toss the ideas in his mind, getting nowhere. Finally, he said he "must sleep on it."[25]

Over the next few weeks, Bohr calmed down enough to

write his rebuttal to the Einstein-Podolsky-Rosen paper. Three months of hard work later, Bohr finally submitted his response to Einstein and his colleagues to the same journal that had published the EPR paper, *Physical Review.* He wrote, in part (the italics are his): "*We are in the freedom of choice offered by [EPR], just concerned with a discrimination between different experimental procedures which allow of the unambiguous use of complementary classical concepts.*" But not all physicists saw the situation this way. Erwin Schrödinger, whose theory was challenged by EPR, told Einstein: "You have publicly caught dogmatic quantum mechanics by its throat." Most scientists were either convinced by Bohr's reply to EPR, or else thought that the controversy was philosophical rather than physical, since experimental results were not in question, and hence beyond their concern. Three decades later, Bell's theorem would undermine this point of view.

WHAT DOES THE EPR PAPER SAY?

According to Einstein, Podolsky, and Rosen, any attribute of a physical system that can be predicted accurately without disturbing the system is an *element of physical reality*.

Furthermore, EPR argue, a *complete* description of the physical system under study must embody all the elements of physical reality that are associated with the system.

Now, Einstein's example (essentially the same one he told Rosenfeld two years earlier) of two particles that are linked together shows that the position and momentum of a given particle can be obtained by the appropriate measurements taken of *another* particle without disturbing its "twin." Thus both attributes of the twin are elements of physical reality.

Since quantum mechanics does not allow both to enter the description of the particle, the theory is incomplete.

The EPR paper (along with Bell's theorem, which followed it) is one of the most important papers in twentieth-century science. "*If, without in any way disturbing a system,*" it declares, "*we can predict with certainty (i.e., with probability equal to unity) the value of a physical quantity, then there exists an element of physical reality corresponding to this physical quantity.* It seems to us that this criterion, while far from exhausting all possible ways of recognizing a physical reality, at least provides us with one such way, whenever the conditions set down in it occur." [26]

EPR then embark on a description of entangled states. These entangled states are complicated, because they use both position and momentum of two particles that have interacted in the past and thus are correlated. Their argument is basically a description of quantum entanglement for position and momentum. Following this description, EPR conclude:

"Thus, by measuring either A or B we are in a position to predict with certainty, and without in any way disturbing the second system, either the value of the quantity P or the value of the quantity Q. In accordance with our criterion for reality, in the first case we must consider the quantity P as being an element of reality, in the second case the quantity Q is an element of reality. But as we have seen, both wave functions belong to the same reality. Previously we proved that either (1) the quantum-mechanical description of reality given by the wave function is not complete or (2) when the operators corresponding to the two physical quantities do not com-

mute the two quantities cannot have simultaneous reality. . . .
We are thus forced to conclude that the quantum-mechanical description of physical reality given by wave functions is
not complete."

What Einstein and his colleagues did was to make what
seems like a very reasonable assumption, the assumption of
locality. What happens in one place does not immediately
affect what happens in another place. EPR say: "*If, without
in any way disturbing a system, we can predict with certainty
(i.e., with probability equal to unity) the value of a physical
quantity, then there exists an element of physical reality corresponding to this physical quantity.*" This condition is satisfied when a measurement of position is made on particle 1
and also when a measurement of momentum is made of the
same particle. In each case, we can predict with certainty the
position (or momentum) of the *other* particle. This permits us
the inference of the existence of an element of physical reality. Now, since particle 2 is unaffected (they assume) by what
is done to particle 1, and the element of reality—the position—of this particle is inferred in one case, and of momentum in the other, both position and momentum are elements
of physical reality of particle 2. Thus the EPR "paradox."
We have two particles that are related to each other. We measure one and we know about the other. Thus, the theory that
allows us to do that is incomplete.

In his response, Bohr said: "The trend of their [EPR] argu-
mentation, however, does not seem to me adequately to meet
the actual situation with which we are faced in atomic
physics." He argued that the EPR "paradox" did not present
a practical challenge to the application of quantum theory to

real physical problems. Most physicists seemed to buy his arguments.

Einstein came back to the problem of EPR in articles written in 1948 and 1949, but he spent most of the remaining time until his death in 1955 trying, unsuccessfully, to develop a unified theory of physics. He never did come around to believing in a dice-playing God—he never did believe that quantum mechanics with its probabilistic character was a complete theory. There was something missing from the theory, he thought, some missing variables that could explain the elements of reality better. The conundrum remained: Two particles were associated with each other—twins produced by the same process—forever remaining interlinked, their wave function unfactorable into two separate components. Whatever happened to one particle would thus immediately affect the other particle, wherever in the universe it may be. Einstein called this "Spooky action at a distance."

Bohr never forgot his arguments with Einstein. He talked about them until the day of his death in 1962. Bohr fought hard to have quantum theory accepted by the world of science. He countered every attack on the theory as if it had been a personal one. Most physicists thought that Bohr had finally settled the issue of quantum theory and EPR. But two decades later Einstein's argument was to be revived and improved by another physicist.

12

Bohm and Aharonov

"The most fundamental theory now available is probabilistic in form, and not deterministic."

—David Bohm

David Bohm was born in 1917, and studied at the University of Pennsylvania, and later at the University of California at Berkeley. He was a student of Robert Oppenheimer until Oppenheimer left Berkeley to head the Manhattan Project. Bohm finished his doctorate at Berkeley and then accepted a position at Princeton University.

At Princeton, David Bohm worked on the philosophy of quantum mechanics, and in 1952 made a breakthrough in our understanding of the EPR problem. Bohm changed the setting of Einstein's challenge to quantum theory—the EPR paper—in a way that made the issues involved in the "paradox" much clearer, more concise, and easier to understand. Instead of using momentum and position—two elements—in the EPR preparation, Bohm changed the thought experiment to one involving two particles with one variable of interest instead of two, the physical element of interest being the *spin* associated with each of the two particles along a particular direction. In the Bohm formulation, as in the original EPR arrangement, both particles can be localized at a distance from each other, so that spin measurements on each are separated in space and time without a direct effect on each other.

Some particles, electrons for example, have a spin associated with them. The spin can be measured independently in any direction the experimenter may choose. Whichever axis is chosen, the experimenter gets an answer of either: "spin up," or "spin down" when a spin measurement is made. When two particles are entangled with each other in what is called a *singlet* state, in which the total spin must be zero, their spin is inexorably linked: if one particle shows spin up, the other will show spin down. We don't know what the spin is, and according to quantum theory the spin is not a definite property until it is measured (or otherwise actualized). Two particles fly off from the same source that made them entangled, and they move away from each other. Particle "A" is measured some time later by Alice, who arbitrarily chooses to measure the particle's spin in, say, the x-direction. According to quantum theory, once particle "A" reveals a spin of "up" in the x-direction, particle "B," if measured by Bob in the x-direction will reveal a "down" spin. The same anti-correlation holds if Alice and Bob choose to measure spin in any other direction, say, the y-direction. (One needs two such directions in order to make the EPR argument using spins.)

In the Bohm version of the EPR thought experiment, two entangled particles are emitted. Once the spin of one of them is measured, and is found to be "up," the spin of the other one must be "down," and this must be so for all directions, for example, both x and y. According to quantum mechanics, the value of the spin in different directions does not have simultaneous reality. But EPR's argument was that all of these are real. Bohm's alteration of the EPR thought experiment simplified the analysis greatly. The Bohm version of the EPR thought experiment is shown below.

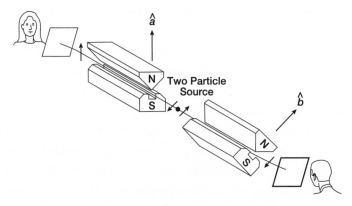

In 1949, Bohm was investigated by the House Committee on Un-American Activities, during the heart of the McCarthy era. Bohm refused to answer questions, but was not charged. However, he lost his position at Princeton University, and as a consequence left the United States to take up a position in Sao Paolo, Brazil. From there he moved for a while to Israel, and then to England, where he became a professor of theoretical physics at the University of London. Bohm continued to work in the foundations of quantum theory, and his discoveries led to an alternative to the "orthodox," Copenhagen view of the discipline.

In 1957, Bohm and Yakir Aharonov of the Technion in

Haifa, Israel, wrote a paper recalling and describing the experiment of Wu and Shaknov that exhibited the spin correlations of Bohm's version of the EPR paradox. The paper argued against the view that perhaps the particles are not really entangled or that the quantum entanglement of particles might dissipate with distance. All the experiments that have been conducted since then confirm this view: the entanglement of particles is real and does not dissipate with increasing separation.

In 1959, Bohm and Aharonov discovered what is now called the Aharonov-Bohm effect, which made them both famous. The Aharonov-Bohm effect is a mysterious phenomenon, which, like entanglement, possesses a non-local character. Bohm and Aharonov discovered a phase shift in electron interference due to an electromagnetic field that has zero field strength along the path of the electron. What this means is that even if we have a cylinder within which there is an electromagnetic field, but the field is limited to the interior of the cylinder, an electron passing outside the cylinder will *still* feel the effects of the electromagnetic field. Thus, an electron that passes outside the cylinder containing the magnetic field will still—mysteriously—be affected by the field *inside* the cylinder. This is demonstrated in the figure below.

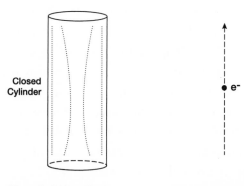

Closed Cylinder

Magnetic field confined to
the insides of the cylinder

The electron feels the effects
of the magnetic field confined
to the cylinder

Like other mysteries of quantum mechanics, no one really understands "why" this happens. The effect is similar to entanglement in the sense that it is non-local. Bohm and Aharonov deduced this effect from theoretical, mathematical considerations. Years later, the Aharonov-Bohm effect was verified experimentally.

Bohm's contributions to our understanding of quantum theory and entanglement are important. His version of the EPR thought experiment would be the one most often used by experimentalists and theorists studying entanglement in the following decades.

In addition, an important requirement for tests of the EPR paradox was laid out by Bohm and Aharonov in 1957. They claimed that in order to find out whether the EPR particles behaved in the way Einstein and his colleagues found objectionable, one would have to use a delayed-choice mechanism. That is, an experimenter would have to choose which spin direction to measure in the experiment only *after* the particles are in flight. Only this design would ensure that one particle, or the experimental apparatus, does not signal to the other particle what is going on. This requirement would be emphasized by John Bell, whose theorem would change our perceptions of reality. An important experimenter would add this requirement to his tests of Bell's theorem, helping establish the fact that entanglement of particles that are remote from each other is a real physical phenomenon.

Niels Bohr and
Albert Einstein at
the 1930 Solvay
Conference.
*Courtesy the Niels Bohr
Archive, Copenhagen.*

Niels Bohr and Werner
Heisenberg in the
Tyrol, 1932.
*Courtesy the Niels Bohr
Archive, Copenhagen.*

Heisenberg and Bohr at the 1936 Copenhagen Conference.
Courtesy the Niels Bohr Archive, Copenhagen.

Niels Bohr with Max Planck, in Copenhagen, 1930.
Courtesy the Niels Bohr Archive, Copenhagen. 1921.

Max Planck in 1921.
Courtesy the Niels Bohr Archive, Copenhagen.

Erwin Schrödinger.
Courtesy the Niels Bohr Archive, Copenhagen.

John Bell.
Courtesy Mary Ross Bell.

(R to L) D. Greenberger, M. Horne, and A. Zeilinger in front of the GHZ experimental design at Anton Zeilinger's lab in Vienna.
Courtesy Anton Zeilinger.

P.K. Aravind.
Courtesy Amir Aczel.

Alain Aspect in his office in Orsay, France.
Courtesy Amir Aczel.

Carol and Michael Horne, Anton and Elisabeth Zeilinger, in Cambridge, MA, in 2001.
Courtesy Amir Aczel.

Abner Shimony.
Courtesy Amir Aczel.

John Archibald Wheeler (right) with the author on the deck
of Wheeler's house in Maine.

Courtesy Debra Gross Aczel.

13

John Bell's Theorem

"For me, then, this is the real problem with quantum theory: the apparently essential conflict between any sharp formulation and fundamental relativity. It may be that a real synthesis of quantum and relativity theories requires not just technical developments but radical conceptual renewal."

—John Bell

John S. Bell, a redheaded, freckled man who was quiet, polite, and introspective, was born in Belfast, Northern Ireland, in 1928 to a working-class family whose members were blacksmiths and farmers. His parents were John and Annie Bell, both of whose families had lived in Northern Ireland for generations. John's middle name, Stewart, was the Scottish family name of John's mother, and at home John was called Stewart until the time he went to college. The Bell family was Anglican (members of the Church of Ireland), but John cultivated friendships that went beyond religion or ethnicity, and many of his friends were members of the Catholic community. Bell's parents were not rich, but they valued education. They worked hard to save enough money to send John to school, even though his siblings left school early to begin work. Eventually, his two brothers educated them-

selves, and one became a professor and the other a successful businessman.

When he was 11, John, who read extensively, decided that
he wanted to become a scientist. He succeeded very well in
the entrance exams for secondary education, but unfortunately his family could not afford to send him to a school
with an emphasis on science, and John had to content himself with admission to the Belfast Technical High School,
where he was educated both academically and in practical
areas. He graduated in 1944 at the age of 16, and found a job
as a technical assistant in the physics department of Queen's
University in Belfast. There, he worked under the supervision of Professor Karl Emeleus, who recognized his assistant's great talent in science and lent him books and even
allowed him to attend freshman courses without being for-
mally enrolled at the university.

After a year as a technician, John was accepted at the university as a student and was awarded a modest scholarship,
which allowed him to pursue a degree in physics. He graduated in 1948 with a degree in experimental physics, and
stayed another year, at the end of which he was awarded a
second bachelor's degree, this time in mathematical physics.
John was fortunate to study with the physicist Paul Ewald, a
gifted German refugee, who was a pioneer in the area of X-
ray crystallography. John excelled in physics, but was
unhappy with the way the quantum theory was explained at
the university. His deep mind understood that there were
some mysteries in this theory that had not been addressed in
the classroom. He did not know, at the time, that these unex-
plained ideas were not understood by anyone, and that it

would be his own work that in time would shed light on these problems.

After working for some time at a physics laboratory at Queen's College in Belfast, Bell entered the University of Birmingham, where he received his Ph.D. in physics in 1956. He specialized in nuclear physics and quantum field theory, and after receiving his degree he worked for several years at the British Atomic Energy Agency.

While doing research on accelerator physics at Malvern in Britain, John met Mary Ross, a fellow accelerator physicist. They were married in 1954, and pursued careers together, often working on the same projects. After they both had earned their doctoral degrees (she received hers in mathematical physics from the University of Glasgow) and worked for several years at Harwell for the British nuclear establishment, they both became disenchanted with the direction the nuclear research center was taking; they resigned their tenured positions at Harwell to assume non-tenured posts at the European Center for Nuclear Research (CERN) in Geneva. There, John worked at the Theory Division, and Mary was a member of the Accelerator Research Group.

Everyone who knew him was struck by John Bell's brilliance, intellectual honesty, and personal modesty. He published many papers and wrote many important internal memos, and it was clear to everyone who knew him that his was one of the greatest minds of the era. Bell had three separate careers: one was the study of the particle accelerators with which he worked; another was the theoretical particle physics he did at CERN; and the third career—the one which ultimately made his name famous beyond the community of physicists—was in the fundamental concepts of quantum mechanics. At conferences organized around him, people congregated who were in the three disciplines John followed, but did not know of each other. Apparently he kept his three careers separate, so people in one discipline did not know he was involved in the other two.

John Bell's working hours at CERN were devoted almost exclusively to theoretical particle physics and accelerator design, so that only his spare time at home could be used to pursue what he called his "hobby"—exploring the basic elements of the quantum theory. In 1963, he took a year's leave from CERN and spent it at Stanford, the University of Wisconsin, and Brandeis University. It was during this year abroad that John began to address the problems at the heart of quantum theory in a serious way. He continued his work on these issues after returning to CERN in 1964, but was careful to keep his involvement with the quantum theory separate from his "main" career at CERN doing particle and accelerator research. The reason was that John Bell had understood early on in his career the serious pitfalls in the quantum theory. While on leave in the U.S., Bell made a breakthrough that told him that John von Neumann had made an error with his assumptions about quantum theory, but, in Bell's words, "I walked away from the problem."

There was no question in anyone's mind that John von Neumann was a superb mathematician—probably a genius. And Bell had no issue with von Neumann's mathematics. It was the interface between mathematics and physics that gave him trouble. In his groundbreaking book on the foundations of quantum theory, von Neumann had made one assumption—which was essential to what came after it—that did not make good physical sense, as John Bell saw it. Von Neumann assumed in his work on quantum theory that the expected value (the probability-weighted average) of the sum of several observable quantities was equal to the sum of the expected values of the separate observable quantities. [Mathematically, for observable quantities A, B, C, $\ldots$, and expectation operator $E(\ )$, von Neumann thought it was natural to take: $E(A + B + C + \ldots) = E(A) + E(B) + E(C) + \ldots$.] John Bell knew that this innocuous-looking assumption was not physically defensible when the observables A, B, C, $\ldots$ are represented by operators that do not necessarily commute with one another. Put roughly in a non-mathematical language, von Neumann had somehow abandoned the uncertainty principle and its consequences, since non-commuting operators cannot be measured at once without loss in precision due to the uncertainty principle.

John Bell wrote his first important paper on quantum fundamentals, which was published as his second paper in this area in 1966 (a later, related paper, which we will discuss soon, being published first). In this paper, "On the Problem of Hidden Variables in Quantum Mechanics," he addressed the error in von Neumann's work as well as similar difficulties with the works of Jauch and Piron, and Andrew Gleason.

Gleason is a mathematician as renowned as von Neumann was. He is a professor at Harvard University who made his name solving one of Hilbert's famous problems. In 1957, Andrew Gleason wrote a paper about projection operators in Hilbert space. Unbeknownst to Bell, Gleason's theorem was relevant to the problem of hidden variables in quantum mechanics. Josef Jauch, who lived for a time in Geneva, where John and Mary Bell lived, brought Gleason's theorem to the attention of John Bell while he was in the process of researching his paper on hidden variables. Gleason's theorem has a certain generality and it is not aimed at solving problems in the quantum theory—it was proved by a pure mathematician with interest in mathematics and not physics. The theorem, however, has a remarkable corollary with important implications to quantum mechanics. The corollary of Gleason's theorem implies that no system associated quantum-mechanically with a Hilbert space with dimension three or greater can admit a dispersion-free state. Bell noticed, however, that if one weakens Gleason's premises, then there is a possibility of a more general kind of hidden-variables theory, a class of theories that today are known as "contextual" hidden-variables theories. Thus there was a loophole if one tried to use Gleason's theorem within the context of the EPR idea.

Dispersion-free states are states that can have precisely measured values. They have no variation, no dispersion, no uncertainty. If dispersion-free states do exist, then the precision they entail comes from some missing, hidden variables, because quantum theory admits an uncertainty principle. Thus to get away from the remaining uncertainty inherent in quantum mechanics in order to achieve these precise, dispersion-free states, one would have to use hidden variables.

Bell didn't understand Gleason's proof of the corollary of his theorem, so he came up with his own proof that showed that except for the unimportant case of a two-dimensional Hilbert space, there are no dispersion-free states, hence no hidden variables. In the case of von Neumann, Bell proved that the assumption used by von Neumann was inappropriate and hence that his results were questionable. Having revived the argument about whether hidden variables exist in the quantum theory, Bell went a step further: he attacked the problem of EPR and entanglement.

Bell had read the 1935 paper by Einstein and his two colleagues, Podolsky and Rosen (EPR), which was published 30 years earlier as a challenge to the quantum theory. Bohr and others had responded to the paper, and everyone else in physics believed that the issue had been closed and that Einstein was shown to have been wrong. But Bell thought otherwise.

John Bell recognized an immense truth about the old EPR argument: he *knew* that Einstein and his colleagues were actually correct. The "EPR Paradox," as everyone had called it, was not a paradox at all. What Einstein and his colleagues found was something crucial to our understanding of the workings of the universe. But it wasn't the claim that the quantum theory was incomplete—it was that quantum mechanics *and* Einstein's insistence on realism and locality could not both be right. If the quantum theory was right, locality was not; and if we insist on locality, then there is something wrong with the quantum theory as a description of the world of the very small. Bell wrote this conclusion in the form of a deep mathematical theorem, which contained certain inequalities. He suggested that if his inequalities could be *violated* by the results of experimental tests, such a violation would provide evidence in favor of quantum mechanics, and against Einstein's common-sense assumption of local realism. If the inequalities were preserved, it would prove that the quantum theory, in turn, was wrong and that locality—in the sense of Einstein—was the right viewpoint. More precisely, it is possible to violate *both* Bell's inequalities and the predictions of quantum mechanics, but it is impossible to obey both Bell's inequalities and the predictions of quantum mechanics for certain quantum states.

John Bell wrote two groundbreaking papers. The first paper analyzed the idea of von Neumann and others about the existence of *hidden variables*, which should be found and added to the quantum theory in order to render it "complete," as Einstein and his colleagues had demanded. In the paper, John Bell proved that von Neumann's and others' theorems proving the impossibility of the existence of hidden variables in quantum mechanics were all flawed. Then Bell proved his own theorem, establishing, indeed, that hidden variables could not exist. Because of a delay in publication, this first important paper by Bell was published in 1966, after the appearance of his second paper. The second paper, published in 1964, was titled "On the Einstein-Podolsky-Rosen Paradox." This paper included the seminal "Bell's Theorem," which changed the way we think about quantum phenomena.

Bell used a particular form of the EPR paradox, one that had been refined into an easier form by David Bohm. He looked at the case in which two entangled spin-1/2 particles in the singlet state are emitted from a common source, and analyzed what happens in such an experiment.

In the paper, Bell said that the EPR paradox has been advanced as an argument that quantum theory could not be complete and must be supplemented by additional variables.

Such additional variables, according to EPR, would restore to quantum mechanics its missing notions of causality and locality. In a note, Bell quoted Einstein:[27]

> But on one supposition we should, in my opinion, absolutely hold fast: the real factual situation of the system S_2 is independent of what is done with the system S_1, which is spatially separated from the former.

Bell stated that in his paper he would show mathematically that Einstein's ideas about causality and locality are incompatible with the predictions of quantum mechanics. He further stated that it was the requirement of locality—that the result of a measurement on one system be unaffected by operations on a distant system with which it has interacted in the past—that creates the essential difficulty. Bell's paper presents a *theorem of alternatives*: either local hidden variables are right, or quantum mechanics is right, but not both. And if quantum mechanics is the correct description of the microworld, then non-locality is an important feature of this world.

Bell developed his remarkable theorem by first assuming that there *is* some way of supplementing quantum mechanics with a hidden-variable structure, as Einstein would have demanded. The hidden variables thus carry the missing information. The particles are endowed with an instruction set that tells them, beforehand, what to do in each eventuality, i.e., in each choice of the axis with respect to which the spin might be measured. Using this assumption, Bell obtained a contradiction, which showed that quantum mechanics *could not* be supplemented with any hidden-variables scheme.

Bell's theorem puts forward an inequality. The inequality compares the sum, denoted by S, of possible results of the experiment—outcomes on the detector held by Alice, and the one held by Bob.

Bell's inequality is: $-2 < S < 2$

The inequality is shown below.

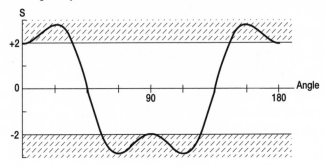

According to Bell's theorem, if the inequality above is *violated*, that is, the sum of the particular responses for Alice and Bob is greater than two or less than minus two, as a result of some actual experiment with entangled particles or photons, that result constitutes *evidence of non-locality*, meaning that something that happens to one particle does affect, instantaneously, what happens to the second particle, no matter how far it may be from the first one. What remained was for experimentalists to look for such results.

There was a problem here, however. Bell derived his inequality from a locality assumption by using a special hypothesis. He assumed that the hidden variables theory agrees exactly with the quantum-mechanical prediction for the two particles in the singlet state, i.e., that along any axis, the spin of particle 1 is opposite to that of particle 2 along the same axis. Hence, if the experimental values agree with the quantum-mechanical prediction of the quantity in Bell's inequality, this finding would not imply the falsity of the locality assumption unless there is evidence that Bell's special assumption is correct, and such evidence is very hard to obtain in practice. This problem would constitute a barrier to definitive experimental testing. But Clauser, Horne, Shimony, and Holt would later derive an improvement that would solve this technical problem and enable actual physical testing using Bell's theorem.

At any rate, the conclusion from Bell's theorem was that hidden variables and a locality assumption had no place within the quantum theory, which was incompatible with such assumptions. Bell's theorem was thus a very powerful theoretical result in physics.

"Do you know why it was Bell, rather than anyone else, who took up the EPR paradox and proved a theorem establishing that non-locality and quantum theory go together?" Abner Shimony asked me. "It was clear to everyone who knew him that it had to be John Bell," he continued. "Bell was a unique individual. He was curious, tenacious, and courageous. He had a stronger character than all of them. He took on John von Neumann—one of the most famous mathematicians of the century—and with no hesitation showed that von Neumann's assumption was wrong. Then he took on Einstein."

Einstein and his colleagues found entanglement between spatially well-separated systems unbelievable. Why would something occurring at one place affect instantaneously something at a different location? But John Bell could see beyond Einstein's intuition and prove the theorem that would inspire experiments to establish that entanglement was a real phenomenon. Bell antecedently agreed with Einstein, but left it for experiment to test whether Einstein's belief about locality was correct.

John Bell died unexpectedly in 1990, at the age of 62, from a cerebral hemorrhage. His death was a great loss to the physics community. Bell had continued to be active to his last days, writing and lecturing extensively on quantum mechanics, the EPR thought experiment, and his own theorem. In fact, physicists today continue to look to Bell's theorem, with its deep implications about the nature of space-time and the foundations of the quantum, as they have over the past three decades. Experiments connected to the theorem have almost all provided overwhelming support for the quantum theory and the reality of entanglement and nonlocality.

14

The Dream of Clauser, Horne, and Shimony

"Our understanding of quantum mechanics is troubled by the problem of measurement and the problem of nonlocality. . . It seems to me unlikely that either problem can be solved without a solution to the other, and therefore without a deep adjustment of space-time theory and quantum mechanics to each other."

—Abner Shimony

*A*bner Shimony comes from a rabbinical Jewish family. His ancestors were among the very few families to have lived continuously in Jerusalem for many generations, and his great grandfather was the chief *shochet* (overseer of kosher slaughtering) of Jerusalem. Abner was born in Columbus, Ohio, in 1928, and grew up in Memphis, Tennessee. From an early age, Abner exhibited a keen intellectual curiosity. As an undergraduate, Abner went to Yale University to study philosophy and mathematics from 1944 to 1948, when he received his bachelor's degree. He read much philosophy, including Alfred North Whitehead, Charles S. Pierce, and Kurt Gödel. While at Yale, he also became interested in the foundations of mathematics.

Shimony continued his studies at the University of Chicago, earning his Master's degree in philosophy, and then went to Yale to do doctoral work in philosophy, earning his Ph.D. in 1953. While at the University of Chicago, Abner studied philosophy with the renowned central figure of the Vienna Circle, an elite European philosophical club, Rudolph Carnap, who later became his informal advisor when Abner was writing his doctoral dissertation at Yale on inductive logic. Carnap seemed baffled by the fact that despite Abner's interest in mathematical logic and theoretical physics he proclaimed himself a metaphysician. This was an appropriate field of interest for him, since he would make his great mark both on physics and on philosophy when he would probe the metaphysical aspects of the concept of entanglement, which would become Abner's obsession and lifelong pursuit within a few years.

At Princeton, Abner met another philosopher with close contacts with the Vienna Circle: the legendary Kurt Gödel. Abner was impressed with the supreme mind that devised the famous incompleteness theorems and proved difficult facts about the continuum hypothesis. Shortly afterwards, Abner decided that he really wasn't that interested in the foundations of mathematics and turned his attention to physics and philosophy. He had become very interested in the philosophical foundations of physics, and so he studied physics and received his Ph.D. in 1962. His dissertation was in the area of statistical mechanics. Shimony became attracted to the quantum theory, and was influenced in his thinking by Eugene Wigner and John Archibald Wheeler.

Shimony has always made a serious effort to combine his philosophical and physical interests carefully. He views physics from a fundamental, mathematical and philosophical point of view, which gives him a unique perspective on the entire discipline and its place within human pursuits. In 1960, before obtaining his second doctorate, Shimony joined the philosophy faculty at M.I.T., teaching courses on the philosophy of quantum mechanics. He began to make a name for himself in this area, and after receiving his second doctorate from Princeton, joined the faculty at Boston University, with a joint appointment in physics and philosophy.

As Abner views it, his was not an expected career path—starting out at a prestigious school such as MIT, getting tenure there, and then switching to an untenured position at a somewhat lower-prestige school (tenure did come to him there very quickly, though). But Abner did it because he wanted to follow his heart. MIT had, and still has, a superb physics department; the institute, in fact, boasts a number of Nobel laureates in physics. But Abner was working within the philosophy department. He longed to teach and do research in both physics and philosophy. So he gave up his tenured position at MIT for a joint appointment in the departments of physics and philosophy at Boston University. The new appointment allowed him to pursue his interests. Our understanding of the complex phenomenon of entanglement—both from a physical and a philosophical point of view—owes much to this move Shimony made to Boston University.

In 1963, Abner wrote an important paper on the measurement process in quantum mechanics. A year later, John Bell wrote his own paper that challenged our understanding of the world.

Abner Shimony first encountered the concept of entanglement in 1957. That year, his new adviser at Princeton, Arthur Wightman, gave him a copy of the EPR paper and asked him if, as an exercise, he could find out what was wrong with the EPR argument. Shimony studied the EPR paper, but found no error in it. Once John Bell's theorem became known to physicists several years later, Wightman had to agree: Einstein had made no errors. What Einstein did was to infer the incompleteness of quantum mechanics from the conjunction of three premises: the correctness of certain statistical predictions of quantum mechanics, the sufficient criterion for the existence of an element of reality, and the assumption of locality. Einstein and his colleagues pointed out to us that if we hold on to our belief that whatever happens in one place cannot instantaneously affect what happens at a distant location, then some phenomena predicted by quantum mechanics will be found in contradiction with such assumptions. It was Bell's theorem, at first ignored by the physics community, that brought this contradiction to the surface in a way that could—at least in principle—be physically tested. What Bell showed was that even if all of the premises of EPR were correct, with the consequence that quantum mechanics would have to be completed with hidden variables, no theory using *local* hidden variables (which, of course, was what EPR desired) would agree with all of the statistical predictions of quantum mechanics. This conflict makes a decisive experiment possible, at least in principle. The essence of this idea was already forming in Abner Shimony's mind.

One day in 1968, Abner Shimony found at his doorstep the first doctoral student he was to supervise as a professor at Boston University's physics department. The student was

Michael Horne. Horne came to Boston after receiving his B.A. in physics from the University of Mississippi, and was excited to work with Shimony.

Michael A. Horne was born in Gulfport, Mississippi in 1943. When he was in high school, the Soviet Union launched the first spacecraft, Sputnik. This event, which had such a profound effect on the development of science in America, as well as so many other facets of our lives, also had a decisive impact on Michael Horne's choice of a career path.

Scrambling to come up with a response to the Russian first in space, the United States convened a council of scientists, the Physical Sciences Study Committee, which met at M.I.T. to devise ways to make America more competitive with the Soviet Union in science, especially physics. The thrust of the program was to make the United States superior in education in the exact sciences, and as part of its recommendations, the Committee commissioned physicists to write science books that would help prepare students in the United States to study physics and other sciences. Mike Horne found one of the books written under the auspices of the Committee at a bookstore in Mississippi and devoured it with great excitement. The book was written by I. B. Cohen, a science historian at Harvard, and was titled *The New Physics*. It was about Newton and his "new" physics of the 1700s. Mike found this to be a beautiful book; he got so much out of the book that he ordered the entire series of books, at 95 cents a volume. The Committee was apparently very successful, at least with Michael Horne: based on what he discovered in these books, he decided during his junior year in high school to become a physicist. When he attended the University of Mississippi, he majored in physics.

Mike was aware of the big physics centers in the United States, and his dream was to do graduate studies at one of them. While still an undergraduate student at the University of Mississippi, Mike Horne read the well-known book by Mach about mechanics. The introduction to the English translation in the Dover edition was written by a Boston University physics professor, Robert Cohen. Mike was taken with the book and the introduction, and wondered whether he would someday meet Robert Cohen, so when he applied to Boston University, he inquired in his letter whether Professor Cohen was still there. Years later, after Mike Horne had made his name as a pioneer in the foundations of physics, Robert Cohen confided to him that the fact that he had asked about him did, indeed, make a difference. Apparently Cohen was so flattered, that he urged the rest of the physics faculty at Boston University to accept Horne to the program in 1965.

Michael Horne was attracted to the foundations of physics as soon as he became interested in the science itself. Thus once he was accepted to graduate studies at Boston University, and had done the first two years of graduate work, he started working with Professor Charles Willis in an area in the foundations of statistical physics. Willis was interested in the problem of deriving rules of statistical mechanics from mechanics, and in similar problems. After doing research with Willis for some time, Horne asked some questions that led Willis to believe that his student would benefit from talking with the philosopher of physics at Boston University, Abner Shimony. And so he sent him to meet him.

Shimony gave Horne the two papers by John Bell, which had recently been sent to him by a friend. Abner knew that the papers were extremely important, and that they were probably being overlooked by the majority of the physics community. Realizing that he had in front of him a student with a keen mind and a great interest in the foundations of quantum theory, Abner handed him the two papers and said: "Read these papers and see if we can expand them and propose a real experiment to test what Bell is suggesting here." Horne went home and began to ponder the obscure but deep ideas that had escaped the attention of so many physicists. What Bell was proposing in his paper was very interesting. Bell thought that Einstein's commitment to locality might possibly be refuted by experiment (although he seemed to be hoping that Einstein's view would win). Was it possible to devise an actual experiment that would test whether Einstein's local realism was right, or whether quantum mechanics—with its implications of non-locality—was right instead? Such an experiment would be of immense value to physics.

John F. Clauser was born in 1942 in California, where his father and uncle as well as other family members had all attended and received degrees from Caltech. John's father, Francis Clauser, had received a Ph.D. degree in physics from Caltech, and at home there were always deep discussions about physics. These conversations took place since John was in high school, and so he was steeped in the tradition of discussions about the meaning and mystery of quantum mechanics. His father stressed to John never to simply accept what people told him, but rather to look at the experimental data. This principle would guide John Clauser's career.

John went to Caltech, and there, studying physics, he asked questions. Clauser was influenced by the teachings of the famous American physicist Richard Feynman, who was on the faculty at Caltech and about whom stories and legends always circulated on campus. John's first rigorous introduction to the quantum theory thus took place at Feynman's lectures, which later were written down and published as the famous "Feynman Lectures on Physics." Volume Three of these lectures is devoted to the quantum theory, and it is in the beginning of this volume that Richard Feynman makes his claim that the result of the Young two-slit experiment contains the essential mystery, and the only mystery, of quantum mechanics.

Clauser caught on quickly to what the key elements in the foundations of quantum mechanics were, and some years later, when he decided to test Bell's inequality and the EPR paradox, he mentioned this desire to his former professor. According to Clauser, "Feynman threw me out of his office."

After Caltech, John Clauser did graduate work in experimental physics at Columbia University. He was there in the late 1960s, working under the supervision of Patrick Taddeus on the microwave background radiation discovery, which was later used by cosmologists to support the big bang theory. But despite the importance of the problem, Clauser was attracted to a different area in physics: the foundations of the quantum theory.

In 1967, Clauser was looking through some obscure physics journals at the Goddard Institute for Space Studies and noticed a curious article. Its author was John Bell. Clauser read the article, and immediately realized something that other physicists had not noticed: Bell's article had potentially immense implications about the foundations of the quantum theory. Bell revived the old EPR paradox and exposed its essential elements. Furthermore, taken literally, Bell's theorem presented a way to experimentally test the very essence of quantum mechanics. Since he was familiar with the work of David Bohm and his extension of the EPR idea in his 1957 work, as well as work by de Broglie, Clauser wasn't completely surprised by Bell's theorem. But having been raised as a skeptic, Clauser tried to find a flaw in Bell's argument. He spent much time looking for a counter-example, trying to refute Bell's remarkable theorem. But after spending weeks on the problem Clauser was satisfied that there was nothing wrong with the theorem; Bell was right. It was now time to make use of the theorem, and to test the very foundations of the quantum world.

Bell's paper was clear to Clauser in every respect save the experimental aspects of the predictions of the theory, which made the cautious Clauser decide to dig through the physics literature looking for experiments that may have been overlooked by Bell, and which might shed light on the problem addressed by the theorem. The only thing Clauser could find, however, was the Wu and Shaknov experiment on positronium emission (the release of two high-energy photons as a result of an electron and a positron annihilating each other) from 1949, which did not fully address the correlation problem. Bell's paper did not provide a clear way for experimentalists to conduct an experiment along the lines of the paper.

Since Bell was clearly a theorist, he assumed—as theorists often do—an ideal experimental setup: ideal apparatus that does not exist in the lab, and ideal preparation of the correlated particles. It was time that someone versed both in theory and in experimental physics took over from where Bell had left off, and designed an actual experiment.

Clauser went to talk with Madame Wu at Columbia to ask her about her own experiments on positronium. As Bohm and Aharonov showed in 1957, the two photons produced in such a way are entangled. He asked Madame Wu whether she had measured the correlations between the photons in her experiments. She said that she had not made these measurements. Had she done so, Clauser thought he could have obtained from her the experimental results he needed to test Bell's inequality. (Wu could not have made such measurements because the high-energy photons from positronium annihilation do not give enough information about pair-by-pair polarization correlation to test Bell's inequality, as Horne and Shimony, and Clauser, were about to find out independently.) Wu sent John to speak with her graduate student Len Kasday, who was redoing her positronium experiments from decades earlier. Kasday and Wu's new experiment (done jointly with J. Ullman) eventually did measure these correlations and would be used to test Bell's inequality. Its results, published in 1975, would be used to add to the evidence in favor of quantum mechanics; although in order to measure the correlations, Kasday and Wu had to make strong auxiliary assumptions they could not test, weakening their results. But this would happen years in the future. For now, Clauser knew that the Wu and Shaknov results were useless in testing Bell's inequality, and he had to develop something new.

All on his own, Clauser kept working, pretty much ignoring what was supposed to be his dissertation area on microwave background radiation. But the reaction of fellow physicists was not favorable. It seemed that no one he talked to thought that Bell's inequalities were worth pursuing experimentally. Physicists either thought that such experiments could not produce results, or they thought that Bohr had already won the debate with Einstein thirty years earlier, and that any further attempts to reconcile Einstein's objections with Bohr's answers would be a waste of time. But Clauser persisted. Going over the results of the old Wu-Shaknov experiment, Clauser concluded that something beyond their experimental results was needed in order to test quantum mechanics against the hidden-variables theories in the way Bell's theorem suggested. He kept working on the problem, and in 1969 he finally made a breakthrough, as a result of which he sent an abstract of a paper to be presented at a physics conference, suggesting how an experiment to test Bell's inequality might be carried out. Clauser's abstract was published in the *Bulletin* of the Washington meeting of the American Physical Society in the spring of 1969.

Back in Boston, Abner Shimony and Mike Horne spent much time in late 1968 and early 1969 steadily working to design what they thought would be one of the most important experiments physicists would ever attempt. Their path was very similar to the one taken by Clauser in New York. "The first thing I did after I got the commission from Abner was to look at the Wu and Shaknov results," recalled Mike Horne. Mike understood that the Wu and Shaknov experiment on positronium annihilation should have had some relevance to the problem of Bell's theorem because the two photons emitted by the electron and positron as they annihilate each other had to be entangled. The problem was that these two photons were of very high energy and, as a result, their polarizations were more difficult to measure than those of visible light. To expose the polarization correlations, Wu and Shaknov had scattered the pairs of photons off electrons ("Compton scattering"). According to quantum mechanical formulas, the correlations between polarization directions of the photons are weakly transferred by the Compton effect into correlations of the directions in space of the scattered particles: that is, up–down or right–left or somewhere in between. Mike suspected, as had John Clauser, that this transfer is statistically too weak to ever be useful in a Bell experiment. To prove this once and for all, Mike constructed an explicit mathematical hidden-variables model that fully met the EPR locality and reality demands and yet reproduced exactly the quantum predictions for the joint Compton scattering.

Thus, the experimental results of Wu and Shaknov—or any future refinement of their experiment using Compton scattering—could not be used to discriminate between the two alternatives: local hidden variables (as suggested by Einstein) vs. quantum mechanics. Something completely new had to be designed.

Mike showed Abner his explicit local hidden variables model, and the two of them decided that visible photons were needed for the experiment. Polaroid sheets, calcite prisms, and some other optical devices exist to analyze the polarization direction of visible-light photons. Such a device is shown below.

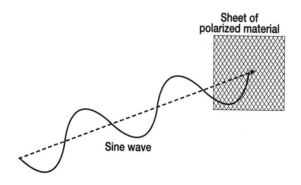

Sheet of polarized material

Sine wave

Abner asked a number of experimentalists for advice on such experiments, and finally learned from an old Princeton classmate, Joseph Snider, then at Harvard, that an optical correlation experiment of the required type had already been conducted at Berkeley by Carl Kocher and Eugene Commins. Abner and Mike soon found out that the Kocher-Commins experiment used polarization angles of zero and ninety degrees only—so their results could not be used to test Bell's inequality, since the intermediate angles were the ones that would offer the determination. Technically, in order to conduct the very sensitive test required to determine between the two alternatives of Bell's theorem (quantum theory versus hidden variables), the experiment had to be carried out at a wide array of such angles. This is shown below.

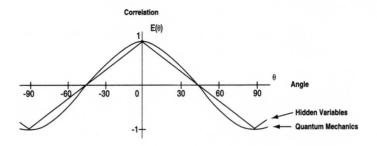

As can be seen from the figure above, the difference between quantum theory and hidden-variable theories is subtle. Only through studying very minutely what happens with pairs of photons as the angle between them changes over a range of values can a researcher detect which of the two theories is correct. Mike and Abner worked on designing the actual experiment with all its requirements so that its results would determine which of the two alternatives was correct: Einstein or quantum mechanics.

They quickly designed a modification of the Kocher-Commins experiment that would allow a physicist to test Bell's inequality under ideal conditions. All an experimentalist would have to do was measure the polarization direction of each photon of an entangled pair along appropriate axes, different from those used by Kocher and Commins. One problem here was the fact that only a few photon pairs would obey the idealized condition of emanation at 180 degrees from each other. So in the next stage, Horne and Shimony relaxed this unrealistic and restrictive assumption and allowed for the collection of photon pairs separated by angles other than 180 degrees. Doing so, however, required a much more complicated calculation to analyze the experimental results. With the help of Richard Holt, a student of Frank Pipkin at Harvard University, who was interested in performing the experiment, Mike Horne was able to calculate the quantum-mechanical predictions for the polarization correlations in this realistic case. Interestingly, these calculations agreed with those performed two years later by Abner Shimony using the quantum-mechanical rules for angular momentum addition.

"This was clearly my best paper on physics," recalled Shimony, when he described to me the paper he and Mike were writing on a design for an experiment to test Bell's inequalities with actual laboratory results in order to see whether nature behaved in a way consistent with the existence of local hidden variables or in accordance with the rules of quantum mechanics. Their proposed experiment would use Bell's magical theorem to determine which of two possibilities was true: Einstein's assertion that quantum mechanics was an incomplete theory, or Bohr's contention that it was complete. In deciding whether the quantum theory was correct, the experiment would also reveal whether, as Einstein feared, there was a possibility of "spooky action at a distance," that is, nonlocal entanglement. Unbeknownst to them, their thoughts at that time were already entangled with those of another physicist, John Clauser, working on the same problem only two hundred miles away.

As part of their preparations, Horne and Shimony spoke with many experts. "We made a nuisance of ourselves," said Shimony. They asked experimentalists about various techniques that would allow them to test the theorem. They had to find an apparatus that would emit pairs of low-energy photons that were entangled with each other, determine a way of measuring their polarizations, calculate the quantum-mechanical predictions for the correlations of these polarizations, and show that the calculated correlations violated Bell's inequality. After many months of work, they finally had a design, and the paper was almost complete. They hoped to present it at the spring meeting of the American Physical Society in Washington, D.C., but missed the deadline for submission. "I thought: What would it matter?" said Shimony, "Who else would be working on such obscure problems? So we passed up on the conference, and prepared to send the paper directly to a journal. Then I got the proceedings for the conference, and discovered the bad news: Someone else had the very same idea." That person was John Clauser.

Abner called Mike early on a Saturday morning. "We've been scooped," he said. The two met the following Monday at the physics department at Boston University, and asked the advice of other physicists: "What should we do?—someone else has done what we have . . . ." Most answered them: "Pretend you don't know about it, and just send the paper to a journal." That didn't seem right to them. Finally, Abner decided to call his own former doctoral adviser at Princeton, Nobel laureate Eugene Wigner. "Just call the man," was Wigner's suggestion, "talk to him about it." So Abner did. He called John Clauser in New York.

While honest and direct, this approach could have had an unpleasant outcome. Scientists tend to be territorial animals, jealously protecting their turf. And since Clauser had already published the abstract of a paper very similar to the one that Horne and Shimony had been working on so hard, he might not have responded well to the newcomers to the same project.

Many people, when finding themselves in such a position, might say: "This is my research project—you got your idea a bit too late!" and hang up the phone. But not John Clauser. To Abner and Mike's great surprise, Clauser's response was positive. "He was thrilled to hear that we were working on the same problem—one that nobody else seemed to care about," Mike Horne told me, recalling that fateful moment.

Actually, Shimony and Horne had a secret weapon at their disposal, which made Clauser even more willing to cooperate with them. The two of them had already lined up a physicist who was ready to conduct the experiment in his lab. This person was Richard Holt, then at Harvard University. In addition to being honestly happy to find two other souls interested in the very same arcane area that attracted him, Clauser knew they could start the experiment, and he wanted to be in on it. Incidentally, Clauser's design of an experiment made the same idealization that Horne and Shimony had originally made—a restriction to photon pairs that are separated by an angle of 180 degrees to each other—and were in the process of eliminating in cooperation with Holt.

Alone, John Clauser would have been left in search of the means to conduct the experiment he sought; and here were Mike Horne, Abner Shimony, and Richard Holt, ready to move forward. He didn't have to think a minute. He was in on it with them.

The four of them, Shimony, Horne, Clauser, and Holt,

began a very fruitful collaboration on the subject, and within a short time produced a groundbreaking paper detailing how an improved experiment could be done to give a definitive answer to Bell's question: Which answer is right, Einstein's local realism, which says that what happens here does not affect what happens elsewhere, or quantum mechanics, which allows for nonlocal entanglement?

The Clauser-Horne-Shimony-Holt (CHSH) paper, published in *Physical Review Letters* in 1969, contained an important theoretical improvement over Bell's pioneering derivation of his inequality. In addition to the existence of a hidden variable that locally determines the outcome of a measurement, Bell had assumed a constraint borrowed from quantum mechanics: that if the same observable quantity is measured in both particles, then the outcomes are strictly correlated. Bell's derivation of his inequality made essential use of this constraint. Clauser, Horne, Shimony, and Holt did away with Bell's restrictive assumption, and thus improved his inequality. The remainder of the paper proposed an extension of the experimental design used by Carl Kocher and Eugene Commins at Berkeley, in which two photons were produced and the correlation between their polarization directions was measured, in a 1966 experiment, without knowledge of Bell's theorem.

Kocher and Commins had used the *atomic cascade* method for producing their correlated photons, and CHSH concurred that this was the right method for their own experiment. Here an atom is excited and emits two photons as it decays two levels down; and the two photons are entangled. The source of the photons was a beam of calcium atoms emanating from a hot oven. The atoms in the beam were bombarded by strong ultraviolet radiation. As a response to this radiation, electrons in the calcium atoms were exited to higher levels, and when they descended again, they released pairs of correlated photons. Such a process is called an atomic cascade because by it an electron cascades down from a high level, through an intermediate level, down to a final level, releasing a photon at each of the two steps down.

Because the initial and the final levels are both states of zero total angular momentum, and angular momentum is a conserved quantity, the emitted photon pair has zero angular momentum, and that is a state of high symmetry and strong polarization correlation between the photons. The idea of such an atomic cascade is demonstrated in the figure below.

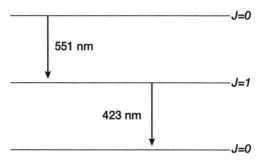

A note at the end of the CHSH paper acknowledged that the paper presents an expansion of the ideas of John Clauser as presented at the spring 1969 meeting of the American Physical Society. Thus a situation that was potentially competitive resulted in a great cooperation, entangling the lives of the four physicists. As John Clauser recalled years later: "In the process of writing this paper, Abner, Mike, and I forged a long lasting friendship that was to spawn many subsequent collaborations."

After receiving his Ph.D. from Columbia, Clauser moved to the University of California at Berkeley to assume a postdoctoral position with the famous physicist Charles Townes, the Nobel laureate who shared in the invention of the laser. Clauser's postdoctoral research project was in the field of radio astronomy, but—as before—he had little interest in anything but the foundations of quantum mechanics. And now, having made the breakthrough into testing Bell's inequality, and with the success of his joint CHSH paper, he had even less patience for anything else. Clauser was ready to perform the actual experiment. The CHSH paper was to be the blueprint for this historic experiment. Fortunately for John, Gene Commins was still at Berkeley. Clauser thus approached Charles Townes and asked him if he would mind if he, Clauser, would spend some time away from radio astronomy trying to perform the CHSH experiment. To his surprise, Townes agreed, and even offered that Clauser spend half his time on the project. Gene Commins was also happy to cooperate on a project that was based on his own past experiment with Kocher, and so he offered Clauser that his own graduate student, Stuart Freedman, would help with the experiment. Back in Boston, Abner and Mike were rooting for him.

Clauser and Freedman began to prepare the apparatus needed for their experiment. Clauser was pushing Freedman to work harder and faster. He knew that back at Harvard, Richard Holt, his coauthor on CHSH, was preparing his own experiment. Freedman was a 25-year-old graduate student with little interest in the foundations of quantum mechanics, but he thought that this should be an interesting experiment. Clauser was desperate to finish the experiment; he knew that Holt and Pipkin at Harvard were moving ahead, and he wanted to be the first to test whether the quantum theory was valid. Deep down, he was betting against the quantum theory, believing that there was a good chance that Einstein's hidden variables were correct and that quantum mechanics would break down on entanglement of photons.

Earlier, while he was still working alone on his paper designing the experiment, Clauser had written to Bell, Bohm, and de Broglie, asking them whether they knew of any similar experiments, and whether they thought such experiments would be important. All had replied that they knew of no such past experiments and that they thought that Clauser's design might be worth pursuing. John Bell was especially enthusiastic—this was the first time that anyone had written him in response to his paper and his theorem. Bell wrote Clauser:[28]

"In view of the general success of quantum mechanics, it is very hard for me to doubt the outcome of such experiments. However, I would prefer these experiments, in which the crucial concepts are very directly tested, to have been done and the results on record. Moreover, there is always the slim chance of an unexpected result, which would shake the world!"

As we will see, there is even a complicated process called *entanglement swapping*, in which two entangled particles swap their mates. In a sense, this is what happened to the people in this grand scientific drama played out across the United States in 1969. Shimony and Horne got entangled with Holt, who was going to conduct an experiment according to their specifications. When they found out about Clauser's own work, they used the fact that Holt was going to do their experiment. As a result, Clauser got entangled with *them*. The four of them created the seminal CHSH paper proposing an important experiment, and Richard Holt got *dis*-entangled with the others and went on to conduct his own experiment. Perhaps this is the reason that in recalling the relationships among them many years later, Clauser mentioned only Horne and Shimony, but not Holt.

Work on performing the experiments proceeded. Bell's enthusiasm, and the support and cooperation from Clauser's new friends in Boston, encouraged Clauser in his quest. Would Bell's inequalities be violated, proving the quantum theory, or would Einstein and his colleagues be the winners and local realism the answer? Clauser, believing in Einstein and local realism, made a bet with Yakir Aharonov of the Technion in Haifa, Israel, with two-to-one odds against quantum theory. Shimony kept an open mind; he would wait and see which theory was correct. Horne believed that quantum mechanics would prevail. He relied on the fact that the quantum theory had been so successful in the past: it never failed to provide extremely accurate predictions in a wide variety of situations.

Clauser and Freedman constructed a source of photons in which calcium atoms were excited into high states. Usually, when the electron in the calcium atom descends back to its normal level, it emits a single photon. But there is a small probability that two photons would be produced, a green one and a violet one. The green and violet photons produced in this way are correlated with each other. The experimental design used by Clauser and Freedman is shown below. Photon pairs produced by the atomic cascade are directed toward polarizers P1 and P2, set at different angles, and then the photons that pass through the polarizers are detected by a pair of detectors, D1 and D2, and finally, a coincidence counter, CC, records the results.

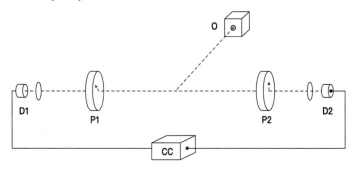

The light signal used in the experiment was weak, and there were many spurious cascades producing non-correlated photons. In fact, for every million pairs of photons, only one pair was detected in coincidence. Later, this flaw would be called the "detection loophole," and the problem would need to be addressed. Because of this low count, it took Clauser and Freedman more than two hundred hours of experimentation to obtain a significant result. But their final result strongly supported the quantum theory and countered Einstein's local realism and hidden variables theories. The Clauser-Freedman result was highly statistically significant. Quantum mechanics beat hidden variables by over five standard deviations. That is, the measured value of S (the quantity used in Bell's inequality) agreed with the prediction of quantum mechanics and was greater than the limit of 2, allowed in the inequality, by five times the standard deviation of the experimental data.

The Clauser-Freedman experiment provided the first definitive confirmation that quantum mechanics is intrinsically non-local. Einstein's realism was dead—quantum mechanics did not involve any "hidden variables." The experiment provided Freedman with his Ph.D. thesis. Clauser and Freedman published the results of their experiment in 1972. The figure below shows their results.

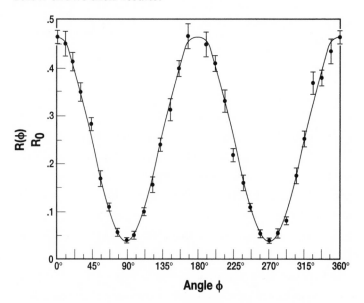

The Clauser-Freedman experiments left some questions unanswered. In particular, the experimental design created a large number of *unobserved* photons, which were produced in order to obtain the entangled pairs. Also, the detectors used were of limited efficiency, and the question arose as to how these limited efficiencies and large numbers of unobserved photons might affect the conclusions. Clauser and Freedmen had done a magnificent job—they provided the best evidence for quantum mechanics and against hidden variable theories. They achieved these results using the best available technology, but this technology was not perfect.

Ironically, while Clauser was a postdoc working for Townes, who had invented lasers, Clauser could not use lasers in his experiment with Freedman, since it was still not known how to do so. Lasers would have helped him and Freedman by enabling them to produce entangled pairs of photons more quickly.

Meanwhile, back at Harvard, Holt and Pipkin had also obtained results. But these were consistent with Einstein and local realism and hidden variables, and against the quantum theory. Since both Holt and Pipkin were believers in the quantum theory, they decided not to publish their results. Instead, they simply waited for the Berkeley team to publish its results, and see what they obtained.

The Holt and Pipkin experiment at Harvard used an isotope of mercury (mercury 200), which exhibits a similar cascade when bombarded by a stream of electrons. Holt and Pipkin's experiment lasted 150 hours, because their experiment, too, suffered from many stray photons. Having seen the Clauser-Freedman results, Holt and Pipkin decided not to go ahead and publish their contrary results in a journal. Instead, in 1973, they distributed an informal preprint of their experimental results to other physicists. Eventually, after others had also come out with experimental results supporting quantum mechanics, Holt and Pipkin concluded that their experiment had suffered from a systematic error of some kind. Although he was no longer working in radio astronomy with the famous Charles Townes, John Clauser managed to stay on at Berkeley as a member of the atomic-beams group headed by Howard Shugart. This allowed him to continue his work. And Clauser, ever the careful experimentalist, decided to revisit his competitors' results and try to replicate them. He was puzzled by their contrary results and wanted to find out the reason for the disagreement. He made only minor modifications of the experimental setup used by Holt and Pipkin, and used a different isotope of mercury (mercury 202) for the atomic cascade. His results, reported in 1976, were again in agreement with the quantum theory and against local hidden-variable theories.

The same year, at Texas A & M University, Ed S. Fry and Randal C. Thompson carried out an experiment with mercury 200, but using a greatly improved design. Because Fry and Thompson excited their atoms with a laser, their light signal was several orders of magnitude more powerful than the signals achieved by the experimenters who did similar work before them. Fry and Thompson were able to obtain their results in only 80 minutes of experimentation. These results supported quantum mechanics and argued against the hidden-variables hypothesis.

In 1978 Abner Shimony was at the University of Geneva in Switzerland. During that year, Abner and John Clauser wrote a joint paper about entanglement, refining their points via long-distance telephone, which surveyed all that was known thus far about the bizarre phenomenon. The article discussed in depth all the experimental findings about entanglement that had been achieved until that year and established that the phenomenon is real. In addition to the experiments mentioned earlier, there were results on Bell's theorem by three other teams that conducted experiments in the 1970s.

One was a group led by G. Faraci, of the University of Catania in Italy. This group, which published in 1974, used high-energy photons (gamma rays) from positronium annihilation (when an electron and a positron annihilate each other). Both Horne-Shimony and Clauser had decided not to do a Bell experiment with photon pairs from positronium annihilation, but the Catania group was able to use data from this kind of experiment by making an additional technical assumption similar to the one made by Kasday, Ullman, and Wu. Doubts about this assumption are responsible for the relative neglect of these experimental results.

Another group, comprised of Kasday, Ullman, and Wu, of Columbia University, which published in 1975, also used positronium annihilation photons. And in 1976, M. Lamehi-Rachti and W. Mittig, of the Saclay Nuclear Research Center, used correlated pairs of protons in the singlet state. The results of these groups agreed with the quantum theory and countered the hidden-variables alternative.

Following the successes in proving the validity of the quantum theory, other theoretical arguments were improved as well. This is usual in science: when the theory advances, the experiments aren't far behind, and when experiments advance, the theory that explains them follows. When one moves forward, the other is not far behind, and once it catches up, it reinforces its symbiote. Bell, Clauser and Horne strengthened the theoretical arguments for testing Einstein's local reality.

They proved a testable inequality, using the assumption of a stochastic (probability-governed) rather than deterministic hidden variables theory. These parallel advances in fundamental physics, all revolving around his remarkable theorem, drew John Bell into the discussion. Clauser, Horne, and Shimony embarked on a years-long exchange of ideas with John Bell.

While all but one of the experiments carried out in the 1970s provided good confirmation of the validity of the quantum theory, it would remain for another scientist, on the other side of the globe, to provide an even better test of Bell's inequality using both laser technology and a greatly improved design that would close an important loophole and thus provide a more complete proof of the mysterious non-local nature of the universe.

In order to really test Einstein's assertion against quantum mechanics, a scientist would also need to account for the possibility—remote and outrageous as it may seem—that, somehow, *signals* may be exchanged between the polarization analyzers at opposite ends of the laboratory. This problem would be addressed by Alain Aspect.

Abner had a dream that he heard a lecture by Alain Aspect, in which Aspect asked whether there is an algorithm—a mechanical decision procedure—for deciding whether a given state of two particles is entangled or not. Abner passed this question on to Wayne Myrvold, an expert on computability in quantum mechanics, who had just had his doctoral thesis accepted by the philosophy department at Boston University. Within two weeks, Myrvold solved the problem. His answer to Aspect's question in Shimony's dream was that no such algorithm is mathematically possible.

15

Alain Aspect

"Bohr had an intuitive feeling that Einstein's position, taken seriously, would conflict with quantum mechanics. But it was Bell's theorem that materialized this contradiction."

—Alain Aspect

Alain Aspect was born in 1947 in a small village in southwestern France, not far from Bordeaux and Perigord, a region in which good food and excellent wines are an integral part of the culture. To this day, Aspect makes his own pâté and keeps his heart healthy by drinking the region's famous red wines. Alain views himself as living proof of what has come to be known as "the French paradox": the fact that the French can eat heavy foods and yet enjoy good cardiovascular health by regularly drinking red wine.

Since early childhood, Alain has been interested in science, especially physics and astronomy. He loved looking at the stars, and he read Jules Vernes's books, especially enjoying *Twenty Thousand Leagues Under the Sea*. He always knew he would become a scientist.

Alain moved to the nearest town to go to school, and after finishing high school he moved to yet a bigger city, Bordeaux, to prepare for the admissions examinations to France's best schools, the renowned *Grandes Ecoles*. He succeeded in passing the examinations and moved to the greatest city of all and the intellectual and academic heart of all Europe: Paris. At the age of 24, Aspect received the graduate degree he calls "my small doctorate," and before continuing to study for his "big doctorate," he took a few years off and volunteered to do social service in Africa. Thus in 1971 he flew to Cameroon.

For three years, under the scorching African sun, Alain Aspect worked hard helping people live better under adverse conditions. But he spent all his spare time reading and studying one of the most complete and deep quantum theory textbooks ever written: *Quantum Mechanics*, by Cohen-Tannoudji, Diu, and Laloë. Alain immersed himself in the study of the bizarre physics of the very small. While working on his degree, he had studied quantum mechanics, but never quite understood the physics, since the courses he took emphasized only the mathematics of differential equations and other mathematical machinery used in advanced physics. Here, in the heart of Africa, the physical concepts themselves were becoming a reality for the young scientist. Aspect began to understand some of the quantum magic that permeates the world of the very small. But of all the strange aspects of the quantum theory, one caught his attention more than all the rest. It was the decades-old proposal by Einstein, Podolsky, and Rosen that was taking on a special meaning to him.

Aspect read the paper by John Bell, then an obscure physicist working at the European Center for Nuclear Research (CERN) in Geneva. And the paper had a profound effect on Aspect, making him decide to devote all his efforts to studying the unexpected implications of Bell's curious theorem. This would lead him down the path to exploring the deepest mysteries of nature. In this, Alain Aspect is similar to Abner Shimony. Both men have a deep—even natural and intuitive—grasp of quantum theory. Each one of them, across the Atlantic from each other, somehow possesses an ability, shared with the late John Bell, of understanding truths that had eluded Albert Einstein.

Like Shimony, Alain Aspect always goes to the origin of a concept or an issue. If he wanted to understand entanglement, Aspect read Schrödinger directly—not an analysis proposed by some later physicists. And if he wanted to understand Einstein's objections to the nascent quantum theory, he searched for and read Einstein's own original papers of the 1920s and 1930s. But surprisingly, beyond the fact that Shimony had a dream in which he saw Aspect make a presentation, leading Shimony to develop an important question, the two men's lives are not entangled. They move in mostly separate circles. While Abner Shimony is an enthusiast, one whose enthusiasm for physics tends to spread to those around him—Horne, Clauser, Greenberger, Zeilinger— spurring them on to greater achievement and discovery, Aspect works differently.

Upon his return from Africa, Alain Aspect devoted himself to a thorough study of quantum theory in his native land. And in fact France was—and still is—an important world center for physics. He found himself in the midst of an elite group of established physicists, from whom he could learn much, and on whom he could test his ideas. The names of the faculty members listed on his dissertation committee read like a *Who's Who* of French science: A. Marechal, Nobel laureate C. Cohen-Tannoudji, B. D'Espagnat, C. Imbert, F. Laloë. The only non-French member on the committee was none other than John Bell himself.

Like Shimony across the Atlantic, Aspect understood Bell's theorem better than most physicists. He was quick to realize the challenge that Bell's remarkable theorem issued to physics and to Einstein's understanding of science. From Aspect's point of view, the essence of the argument between Bohr and Einstein was Einstein's conviction that:

"We must abandon one of the following two assertions: 1. The statistical description of the wave function is complete; or: 2. The real states of two spatially separated objects are independent from one another."[29]

Aspect understood very quickly that it was this assertion by Einstein, as articulated in the EPR paper of 1935, which John Bell's theorem addressed so succinctly and elegantly. Using the EPR setup, Bell offered an actual framework for testing the hypothesis that the quantum theory was incomplete versus the assertion that it was, indeed, complete but included distinctly non-local elements.

Bell's theorem concerns a very general class of *local* theories with hidden, or supplementary, parameters. The assumption is as follows: suppose that the quantum theory is *incomplete* but that Einstein's ideas about locality are preserved. We thus assume that there must be a way to *complete* the quantum description of the world, while preserving

Einstein's requirement that what holds true *here* cannot affect what holds true *there*, unless a signal can be sent from here to there (*and such a signal, by Einstein's own special theory of relativity, could not travel faster than light*). In such a situation, making the theory complete means discovering the hidden variables, and describing these variables that *make* the particles or photons behave in a certain way. Einstein had conjectured that correlations between distant particles are due to the fact that their common preparation endowed them with hidden variables that act locally. These hidden variables are like instruction sheets; and the particles' following the instructions, with no direct correlations between the particles, ensures that their behavior is correlated. If the universe is *local* in its nature (that is, there is no possibility for super-luminal communication or effect, i.e., the world is as Einstein viewed it) then the information that is needed to complete the quantum theory must be conveyed through some pre-programmed hidden variables.

John Bell had demonstrated that any such hidden-variable theory would not be able to reproduce all of the predictions of quantum mechanics, in particular the ones related to the entanglement in Bohm's version of EPR. The conflict between a complete quantum theory and a local hidden variables universe is brought to a clash through Bell's inequality.

Alain Aspect understood a key point. He knew that the quantum theory had by this time enjoyed a tremendous success as a predictive tool in science. He thus felt that the apparent conflict described above and inherent in Bell's theorem and its attendant inequalities could be used, on the contrary, to defeat all local hidden-variables theories. Thus, unlike John Clauser, who before his experiment bet that the quantum theory would be defeated and that locality would win the day, Aspect set out to design his own experiments believing that the quantum theory would be victorious and that locality would be defeated. If his contemplated experiments should succeed, he mused, non-locality would be established as a real phenomenon in the quantum world, and the quantum theory would repel the attack upon its completeness. It is important to note, however, that whatever proclivities Clauser and Aspect may have had concerning the expected outcomes of their respective experiments, each designed an experiment to allow nature to speak without any preexisting bias one way or the other.

Aspect was well aware that Bell's theorem, virtually ignored when it first appeared in the mid-1960s, had become a tool for probing the foundations of the quantum theory. In particular, he knew about Clauser's experiments in California and the involvement of Shimony and Horne in Boston. He was also aware of several inconclusive experiments. Aspect realized, as he later stated in his dissertation and subsequent papers, that the experimental setup used by the physicists whose work came before his was difficult to use. Any imperfection in experimental design had the tendency to destroy the delicate structure that would have brought about the desired conflict between Bell's inequalities and quantum predictions.

The experimenters were looking for outcomes that were very hard to produce. The reason for this was that entanglement is a difficult condition to produce, to maintain, and to measure effectively. And in order to prove a violation of Bell's inequality, which would prove quantum predictions, the experimental design had to be constructed very carefully. Aspect's aim was to produce a superior experimental setup, which would allow him, he hoped, to reproduce Bohm's version of the EPR thought experiment as closely as possible, and allow him to measure the correlations in his data for which quantum mechanics predicts a violation of Bell's inequalities.

Aspect set to work. He built every piece of equipment on his own, working in the basement of the Center for Optical Research of the University of Paris, where he'd been given access to experimental space and apparatus. He built his source of correlated photons, and constructed the arrangement of mirrors, polarization analyzers, and detectors. Aspect considered carefully the thought experiment of EPR. In the version proposed by David Bohm, the phenomenon in question is simpler and Bell's theorem applies: the spins or polarizations of two particles are correlated. In contrast, Einstein's momentum and position framework are more complicated because these two quantities have a continuum of values and Bell's theorem is not directly applicable. After thinking about the problem for a long time, Alain Aspect reached the conclusion that the best way to test the EPR conundrum would be with the use of optical photons, as had been done in the best earlier experiments.

The idea, previously followed by Clauser and Freedman as well as their colleagues in Boston Shimony, Horne, and Holt, was to measure the polarization of photons emitted in correlated pairs. Aspect knew that a number of experiments of this kind had been carried out in the United States between

1972 and 1976. The most recent of these experiments leading to results in support of quantum mechanics, conducted by Fry and Thompson, was carried out using a laser to excite the atoms.

Aspect decided to carry out a series of three main experiments. The first was a single-channel design aimed at replicating the results of his predecessors in a much more precise and convincing way. He would use the same radiative cascade of calcium, in which excited atoms emit photons in correlated pairs. Then, he would conduct an experiment with two channels, as had been proposed by Clauser and Horne to get closer to an ideal experiment. If there is only one channel, then the photons that do not enter it may behave as they do for one of two reasons: either they hit the polarization analyzer but have the wrong polarization to pass through, or they miss the entrance of the analyzer. With two channels, one can restrict attention to the particles that are detected— all of these must have hit the entrance aperture and exited through one channel or the other. Such a methodology helps close the detection loophole. Finally, Aspect would conduct an experiment that was suggested by Bohm and Aharonov in 1957 and articulated by John Bell. Here, the direction of polarization of the analyzers would be set *after* the photons had left their source and are in flight. This is a type of design in which experimenters play devil's advocates. In a sense, the experimenter is saying: "What if one photon or its analyzer *sends a message* to the other photon or its analyzer, informing the other station of the orientation of the analyzer, so that the second photon can adjust itself accordingly?" To prevent such an exchange of information, the experimenter chooses the orientation to be used in the design both *randomly* and *with delay*. Thus, what Alain Aspect was after, was a more definitive test of Bell's inequality—a test whose results could not be doubted by someone who thinks that the analyzers or photons communicate with each other in order to fool the experimenter. It should be noted that, in the thinking of physicists, communication may not be such a bizarre notion, and the intent to fool the experimenter is absent from such thinking. What the physicists are worried about is the fact that in a physical system that has had a chance to reach some equilibrium level, communication by light or heat may transfer effects from one part of the system to another.

In the actual experiment, Aspect had to resort to a signal that was periodic, rather than perfectly random—however, the signal was sent to the analyzers after the photons were in flight. This is the essentially new and important element of his experiments.

Aspect's two-channel, but non-switched arrangement (reprinted by permission from his dissertation) is shown below.

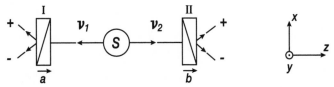

Since Alain knew that Bell's inequality had previously been used to determine which of the two alternatives, quantum mechanics or local realism, was true, he went to Geneva to visit John Bell. He told him that he was planning an experiment that would incorporate a dynamic principle of time-varying polarizers to test for Einstein separability, as Bell himself had suggested in his paper. Bell looked at him and asked: "Are you tenured?" to which Aspect responded by saying he was only a graduate student. Bell stared at him with amazement. "You must be a very courageous graduate student . . . " he muttered.

Aspect began his experiments, and used an atomic beam of calcium as his source of correlated photons. The atoms were excited by a laser. This caused an electron in each atom to move up two levels of energy from its ground state (as was done in previous experiments). When the electron descended two levels down, it sometimes emitted a pair of correlated photons. The energy levels and the entangled photons produced by this method of atomic cascade are shown below.

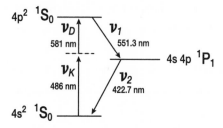

The coincidence rate for the experiment, the rate at which correlated pairs were indeed detected and measured, was several orders of magnitude higher than the rate obtained by Aspect's predecessors. These experiments with a single-channel polarizer led to excellent results: Bell's inequality was violated by nine standard deviations. This means that quantum theory prevailed, no hidden-variables were found to be possible, and nonlocality was inferred to exist for these entangled photons—they respond instantaneously to one another—with an immensely small probability that these conclusions are wrong. This result was very powerful. Next, Aspect carried out his two-channel experiments.

When a photon is blocked by the polarizer in a single-channel design, that photon is lost and there is no way to determine whether it was correlated with another one and how. This is why two-channels were used. What happens here is that if a photon is blocked by the polarizer then it is reflected by it and can still be measured. This increases the coincidence rate of the overall test and makes the experiment much more precise. With this greatly improved scheme of measurement, the results obtained by Aspect were even more precise and convincing. Bell's inequality was violated by more than 40 standard deviations. The evidence in favor of quantum mechanics and non-locality was overwhelming and went far beyond anyone's expectations.

Then came the ultimate test of non-locality, a test of whether a photon could still send a signal to another, versus the quantum-mechanical alternative that non-locality prevails and that the photons—without being able to send signals to one another—react to each other's situation instantaneously. Aspect designed polarizers whose direction in space could be changed at such high speed that the change is made *while the two photons are in flight*. This was achieved in the following way. On each side of the experiment, there were two polarization analyzers, using different orientations. Both were connected to a switch that could rapidly determine to which of the two analyzers to send each photon, and thus which of two possible orientations the photon would encounter. This innovation, in fact, was the greatest of the Aspect experiments, and the one that was widely viewed as the ultimate test of non-locality.

Aspect's third set of experiments, with switching between analyzers while the pairs of photons were in flight, is shown in the figure below.

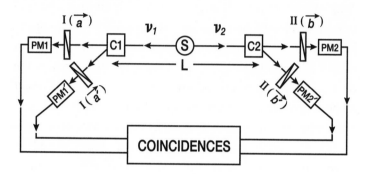

In explaining the design of his third set of experiments, Aspect quoted an important statement by John Bell: "The settings of the instruments are made sufficiently in advance to allow them to reach some mutual rapport by exchange of signals with velocity less than or equal to that of light." In such a case, the result at polarizer I could depend on the orientation, b, of the remote polarizer II, and vice versa. In this case, the locality condition would not hold and could not be tested." The scientists are being very careful here. They play devil's advocate, allowing for the possibility that the polarizers and the photons interact with each other and provide results consistent with local reality. At any rate, when the polarizers in the experiment are fixed, the locality condition is not enforced and so—in the strictest sense—it is not possible to test the EPR idea, which demands local realism, against the quantum theory using Bell's theorem.

In Aspect's lab, each of the polarizers was placed at a distance of 6.5 meters from the source. The total distance between the two polarizers as shown in the diagram above was 13 meters. So, in order to solve the problem and allow for an objective test of "Einstein causality," meaning a test in which the photons and polarizers can't "cheat the experimenter" by sending signals to one another, Aspect had to design an experimental way of *switching* polarizer I between the settings a and a' and polarizer II between its two settings of b and b' in an interval of time that was less than 13 meters divided by the speed of light (about 300,000,000 meters per second), which is about 4.3×10^{-8} seconds (43 nanoseconds). Aspect was able to achieve this goal and to build a device able to respond at such incredible speeds.

In the experimental setup shown in the diagram of Aspect's experiment, the switching is achieved in less than 43 nanoseconds. The switching is done by an acousto-optical device in which light interacts with an ultrasonic standing wave in water. When the wave changes in the transparent water container, the beam of light hitting the water is deflected from one setting to another. In fact, the switching took place at intervals of 6.7 and 13.3 nanoseconds, well below the maximum of 43 nanoseconds.

Aspect's third set of experiments was also successful, and again locality and hidden-variables were defeated in favor of quantum mechanics. Aspect noted that he would have liked to have an experimental setup in which not only the settings are changed while the photons are in flight, but also in which the switching is done purely at random. His design did not provide for randomness, but rather a cyclical change of settings. So, as Anton Zeilinger has pointed out, an extremely clever group of photons and polarizers could—in principle—"learn" the pattern and try to fool the experimenter. This, of course, would be extremely unlikely. Still, Aspect's third set of experiments contained an immensely important dynamic component, which added to the power of his entire set of positive results for quantum mechanics and helped establish non-local entanglement as a real phenomenon.

The figure below shows, as shaded area, the region in which Einstein's locality fails in the experiments.

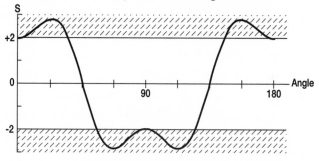

In the following years, still working at the Center of Optics at the University of Paris in Orsay, Aspect went on to conduct other important experiments in quantum physics. Recalling his groundbreaking entanglement experiments of the 1980s, he said: "I am also proud of the fact that, besides doing the experiments, my work also called attention to Bell's theorem. At the time I did the work, this wasn't a popular field."

16

Laser Guns

"[Interference occurs because] one photon must have come from one source and one from the other, but we cannot tell which came from which."

—Leonard Mandel

*F*ollowing the tremendous success of the Aspect experiments, which demonstrated definitively (to most physicists' minds) the reality of entanglement, the study of the phenomenon progressed. While Alain Aspect and his colleagues at Orsay, as well as researchers who had done earlier experiments, used the atomic cascade method for producing entangled states, right after these experiments were concluded, in the early 1980s, experimental physicists began to use a new method. This method, which is still the preferred technique for producing entangled photons today, is called *spontaneous parametric down-conversion* (SPDC for short).

Imagine that there is a transparent crystal sitting on a table, and that someone shines a light on this crystal. At first, you only see the light that comes through the crystal, shining out the other side. But as the intensity of the light is increased, suddenly you see an additional effect: a pale halo that surrounds the crystal. When you look closer, you notice that the faint halo is shimmering with all the colors of the rainbow. This beautiful phenomenon is produced by an interesting physical effect. It turns out that, while most of the light that is shone on the crystal passes through it to the other side, a very small percentage of the light entering the crystal does not go straight through. This minority of photons undergoes a bizarre transformation: each photon that does not go straight through the crystal "breaks down" into *two photons*. Each such photon somehow interacts with the crystal lattice, in a way that is not completely understood by science, and this interaction gives rise to a pair of photons. When the photon undergoes this transformation, the sum of the frequencies of the two resulting photons is equal to the frequency of the original photon. The photons in a pair produced in this way are entangled.

In the down-conversion method of producing entangled photons, scientists use a laser to "pump" the crystal with light. The crystals used for this purpose are special ones that exhibit this property of generating photon pairs. Among the crystals that can be used are lithium iodate and barium borate. Such crystals are known as *non-linear crystal*s. That is because when the crystal lattice atoms are excited, the resulting energy that comes out of the lattice is described by an equation which includes a non-linear (squared) term. The down-conversion method has been used by physicists since 1970. That year, D. C. Burnham and D. L. Weinberg discovered the phenomenon when they examined the nature of secondary light produced when intense laser light passed through a nonlinear crystal, and the crystal seemed suddenly bathed in a weak rainbow of colors. The scientists discovered that most of the light passed through the crystal, but that about one in a hundred-billion photons gave rise to two photons. Because the two resulting photons had frequencies adding up to that of the original single photon (meaning each of them has been bumped down in frequency), physicists named the process *down-conversion*. A single photon was converted downward in its frequency to the lower frequencies of the two resulting photons. But the researchers did not realize that the two photons thus produced were in fact entangled, and that they had just discovered a valuable way of producing entangled photons. These photon pairs are not only entangled in their *polarization*, but also in their *direction*, which is useful for studies involving two-photon interference.

Scientists experimenting with entanglement using the older, atomic cascade method had noticed that there was a collection efficiency loophole. This effect is due to atomic recoil. When the atoms recoil, some of the momentum is lost from consideration. Thus the angles made by the resulting entangled photons were not precisely known, making it difficult to identify by direction which photon is associated with another in an entangled pair. The down-conversion method is much more precise. It is illustrated in the figure below.

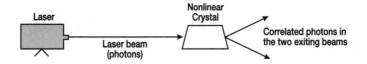

The first scientist to make use of the down-conversion method to study entanglement was Leonard Mandel. Mandel was born in Berlin in 1927, but moved with his family to England while he was a young child. He received a Ph.D. in physics from the University of London in 1951, and became a senior lecturer in physics at Imperial College, the University of London, where he taught until 1964. That year, Mandel was invited to join the physics faculty at the University of Rochester, in New York. In America, Mandel did work on cosmic rays, which entailed climbing to the tops of high mountains with experimental apparatus that could detect and measure these high-energy particles as they passed through Earth's atmosphere. At high altitude there were many more such particles that could be measured than at lower levels. After a number of years of this research, Mandel became fascinated with optics as well as with the quantum theory, which governs the behavior of the particles he had been studying.

In the late 1970s, Leonard Mandel embarked on a series of experiments, some with H. Jeff Kimble, demonstrating quantum effects with laser light. Some of these experiments bounced photons off individual sodium atoms. Some of the experiments dealt with complementarity: the wave-particle duality of light and the quantum-mechanical idea that one of these aspects of light, but not both, can be evidenced through any single experiment. Mandel's experiments demonstrated some of the most striking quantum properties of light. In some experiments, Mandel has shown that if the experimental design merely allowed the experimenter the *possibility* of measurement, that was enough to change the outcome of the experiment from a wave-pattern to particle-like behavior.

In the 1980s, Leonard Mandel and his colleagues began to use the parametric down-conversion technique to produce entangled photons. One of these experiments, whose results were published in 1987 in a paper by R. Ghosh and L. Mandel in the journal *Physical Review Letters* (vol. 59, 1903-5), demonstrated an interesting fact about entanglement. The Ghosh and Mandel experimental design is shown on the next page.

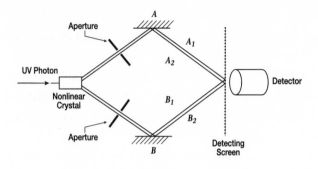

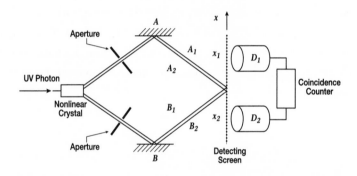

In the experiment above, a nonlinear crystal is pumped by a laser, producing pairs of entangled photons. Since the photon that enters the crystal can produce a pair of photons in any of infinitely many ways (because all that's required is that the sum of the frequencies of the progeny be equal to the frequency of the parent photon), within a certain range of distance on the screen there can be found photons that are entangled with each other.

In the experiment shown in the upper diagram, a single, tiny detector is moved along the screen. Ghosh and Mandel found, surprisingly, that *no* interference is present. Hence a single photon does not exhibit the interference pattern that one would expect based on the old Young double-slit experiment. In the second experiment, shown on the bottom part of the figure above, *two* detectors are used, at separate points on the screen. Again, when each detector was moved along the screen, no interference pattern was exhibited. Ghosh and Mandel then hooked the two detectors to a coincidence counter: a counter that registers a count only if both detectors fire together. Now, when they fixed one of the detectors and moved the other one along the screen, they found that the coincidence counter registered a clear interference pattern similar to the one shown in the Young double-slit experiment.

The reason for this surprising finding is that, while in quantum theory a single photon is shown to travel both paths and to interfere with itself, as exhibited by the Young experiment, with entangled photons the situation is different. An entangled pair of photons constitutes a *single entity* even while they are separate from each other. What happens here is that the entangled two-photon entity is in a superposition of two product states, and thus is the entity that interferes with itself. This is why the interference pattern appears only when we know what happens simultaneously at *two* locations on the screen—that is, when we track the two entangled photons as a single entity—and only in this framework do we find the familiar peaks and valleys of intensity interference, for a pair of photons seen as one element. Here, two distant observers, one placed at each detector, must compare their data in order to see that something is happening—each observer alone sees only a random arrival of photons, with no pattern, and with constant average count rate. This finding demonstrates an important idea about entanglement: that it is not correct to think of entangled particles as separate entities. In some respects, entangled particles do not have their own individual properties but behave as a single entity.

Another kind of experiment was proposed in 1989 by James Franson of Johns Hopkins University. He pointed out that two-particle interference fringes can arise when we don't know *when* the two particles were produced. Raymond Chiao of the University of California at Berkeley and his colleagues have performed an experiment based on Franson's design, and so have Mandel and his colleagues. This kind of experimental arrangement uses a short and a long path in each of two arms, separated by half-silvered mirrors. Which route did each photon take? The entangled photons in the down conversion are produced at the same time, and arrive together. But since we don't know when they were produced, we have a superposition of the long path for both photons *and* the short path for both photons. This produces a *temporal* double-slit arrangement.

Another physicist to make extensive use of the SPDC technique to produce entangled photons was Yanhua Shih of the University of Maryland, who in 1983 began a series of experiments aimed at testing Bell's inequality. His experiments were very precise and led to results in good agreement with quantum mechanics and in violation of Bell's inequality. Shih and his colleagues were able to demonstrate a violation of the Bell inequality to an extent of several *hundred* standard deviations. These results were statistically very significant. Shih's team conducted experiments with delayed-choice setups as well, and here, too, they were able to confirm the agreement with quantum mechanics to very high accuracy.

Shih then studied the effects of a perplexing phenomenon called the quantum eraser. When we can tell, using detectors in an experiment, which of two paths were taken by a photon, no interference pattern appears. Thus, in a "which-path" design, we observe the particle-like nature of light. If the experimental design is such that the experimenter cannot tell which of two paths were taken by a photon, we are in the quantum "both-paths" design. In this case, the photon is viewed as taking both paths simultaneously. Here, an interference pattern can appear and the experiment thus exhibits the wave-nature of light. Recall that by Bohr's principle of complementarity, it is impossible to observe in the same experiment both the wave- and the particle-nature of light.

Shih and his colleagues constructed strange experiments that can "erase" information. Even more stunningly, they used a delayed-choice eraser. Here, an entangled photon pair was produced and injected into a complex system of beam-splitters (half-silvered mirrors that reflect a photon or pass it through with probability one-half). After one photon was already registered, in terms of its position on a screen, the setup was switched randomly such that some of the time the experimenter could tell which path was taken and some of the time not. Thus it could be determined after the first photon hit the screen whether it *had* wave or particle nature when it hit the screen based on what was a fraction of a second later encountered by its twin that was still in flight.

But the most interesting experiment Shih and his colleagues

have performed from the point of view of this book, and also with an eye toward applications in technology, was the *Ghost Image Experiment*. This experiment used one member of each pair of entangled photons to make the other, distant member of the same pair help create a "ghost" image at the distant location.[30] The diagram of this experiment is shown below.

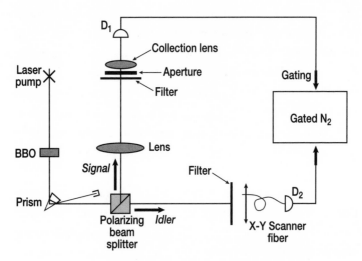

As we see from the figure, a laser pumps a nonlinear crystal (barium borate), producing the SPDC entangled photons, which then go through a prism and on to a beam splitter that splits them based on their polarization direction. Thus, one of each pair of entangled photons goes up, through a lens, and encounters a filter with an aperture. The aperture is in the form of the letters UMBC (for University of Maryland, Baltimore County—Shih's university). Some of the photons are blocked, but the ones that go through the letter-apertures are then collected by a lens and detected by a detector. The first detector is linked to a coincidence counter along with the second—which is collecting the twin photons that go through the filter. These twins went straight through the beam splitter. They hit a filter and a scanning fiber that records their locations on the screen. Only those in coincidence with the twins that went through the UMBC apertures are recorded. They form the ghost image of UMBC on the screen. This ghost image is shown below.

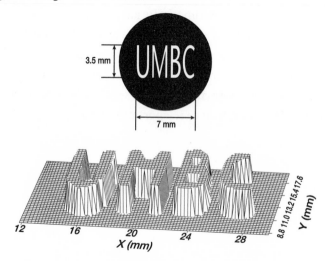

Thus, using entangled photons, the image UMBC was transported to a distant location by twins of photons that went through the letters, providing a dramatic demonstration of an interesting aspect of entanglement. The image is transformed to create the ghost using two elements. First, we have the photons arriving at the screen with the scanning fiber: but not all arriving photons are counted. We communicate with whoever is observing the twins, the photons entangled with the photons arriving at the screen, by using the coincidence counter. We only count screen photons that "double click" with a twin that has passed through the letter-aperture. It is this combination of entanglement with a "classical channel" of information that allows us to create the ghost image.

The next stage in Yanhua Shih's career took him to the most exciting project of all: quantum teleportation. Some basic ideas about teleportation have analogous twins in the ideas of the ghost experiment. In particular, quantum teleportation entails the use of two channels simultaneously: an "EPR channel," meaning a channel of the "action-at-a-distance" of entanglement (which is immediate); and a "classical channel" of information (whose speed is limited by that of light). We will return to teleportation later.

17

Triple Entanglement

"Einstein said that if quantum mechanics were correct then the world would be crazy. Einstein was right—the world is crazy."

—Daniel Greenberger

"Einstein's 'elements of reality' do not exist. No explanation of the beautiful dance among the three particles can be given in terms of an objectively real world. The particles simply do not do what they do because of how they *are*; they do what they do because of quantum magic."

—Michael Horne

"Quantum mechanics is the weirdest invention of mankind, but also one of the most beautiful. And the beauty of the mathematics underlying the quantum theory implies that we have found something very significant."

—Anton Zeilinger

When we last left Mike Horne, he was enjoying the fruits of the success of his work with Abner Shimony, John Clauser, and Richard Holt (CHSH) and the actual demonstration of entanglement by an experiment testing Bell's inequality, with results favoring quantum mechanics, carried out by Clauser and Freedman. The success of CHSH and its attendant experimental demonstrations received wide attention in the physics literature and made scientific news. There were expository articles published in journals that report on new discoveries, and there were new experiments and renewed excitement about the foundations of the strange world of the quantum.

Soon afterwards, Clauser, Shimony, and Horne got involved with the man who started it all: John Bell. The four men began an extensive communication, some appearing as research papers, intended to answer questions and discuss ideas proposed by one side or the other. This fruitful communication resulted in Bell's theorem being based on less-restrictive assumptions, and it also improved our understanding of the amazing phenomenon of entanglement.

In 1975, Mike Horne joined a research group headed by Cliff Shull of M.I.T., which performed experiments on neutrons produced at the M.I.T. nuclear reactor in Cambridge. Mike spent ten years at the reactor, conducting single-particle interference experiments with neutrons. He also met two physicists who would change the course of his career, and whose joint work with him would produce a giant leap in our understanding of entanglement. The two scientists were Daniel Greenberger and Anton Zeilinger. The three of them would write a seminal paper proving that *three* particles could be entangled, and would spend years studying the properties of such entangled triples. When, years later, I asked them whether the three of them were somehow "entangled" themselves, just as the triples of particles they had studied, Anton Zeilinger quickly responded: "Yes, in fact we were so close that when one of us would open his mouth to say something, the others would finish his sentence for him . . ."

Michael Horne's path from two-particle interference studies to one-particle interference research had a good reason behind it. Having done the CHSH work that helped establish entanglement as a key principle in the foundations of quantum mechanics, Mike decided to study further problems in these very foundations. He knew very well the history of the development of ideas in the quantum theory as the discipline evolved. He knew that when Young did his amazing experiment with light in the 1800s and discovered the interference pattern that still puzzles us today, light (and other electromagnetic radiation) was the only microscopic "wave" known. Then, of course, in 1905, Einstein proposed the photon as a solution to the photoelectric effect, showing that light was not only a wave but also a stream of particles. Mike also knew that in 1924, de Broglie "guessed that even particles are waves," as Mike put it, but that "no one at that time could perform a two-slit experiment with electrons, although direct confirmation of de Broglie's waves did come quickly from crystal diffraction of electrons." A quarter century later, in the 1950s, the German physicist Möellenstedt and co-workers did perform the experiment. They showed that these particles, the electrons, display the same wave-nature exhibited by an interference pattern on a screen once they emerge from Young's old double-slit setup.

Then, in the mid-1970s, first Helmut Rauch in Vienna and then Sam Werner in Missouri independently performed what was essentially a double-slit experiment with neutrons. These massive quantum objects exhibit the same interference patterns that we associate with waves as they emerge from the two-slit experimental setup. The two teams, in Vienna and in

Missouri, both used thermal neutrons: neutrons produced by reactions that take place inside a nuclear reactor. These neutrons travel at low speeds (that is, "low" as compared with the speed of light) of about a thousand meters per second, and thus by de Broglie's formula, their associated wavelength is measured in a few angstroms. These very challenging experiments were now possible because of new semiconductor technologies, which made available large, perfect silicon crystals. The scientists used hand-sized silicon crystals to construct interferometers for the thermal neutrons coming from the reactor. As the neutrons interacted with the crystal lattice, the beam of neutrons was first split by diffraction at one slab of the crystal, and then other slabs were used to redirect and eventually recombine the beams to produce the interference pattern.

Mike was very interested in these experiments, which had just been done. He knew that Cliff Shull, one of the pioneers of neutron work in the 1940s (who, in 1994, would receive a Nobel Prize), had a lab at the M.I.T. reactor and was working there performing experiments on thermal neutrons. Mike already had a position teaching physics at Stonehill College, but Stonehill did not have a reactor or a well-known physicist directing exciting new research. So one day in 1975, Mike walked into Cliff Shull's lab at M.I.T. and introduced himself. He mentioned to Shull his work on entanglement with Abner Shimony and John Clauser, and his interest in neutron interference experiments. Then he asked: "Can I play?"

"Take that desk right there," was Shull's answer as he pointed to a desk on the side of the lab. From that day on, for ten years, from 1975 to 1985, every summer, every Christmas vacation, and every Tuesday (the day he didn't teach), Mike Horne spent at Shull's lab at the M.I.T. reactor doing work on neutron diffraction. Two experiments that he found especially attractive had already been performed with neutrons in Vienna and in Missouri. Cliff Shull's group would conduct many more such experiments at M.I.T.

The experiment done by Sam Werner and collaborators at the University of Missouri in 1975 demonstrated directly how neutron two-slit interference is affected by gravity—something that had not been shown before. There had never been a demonstration of the effect of gravity on quantum mechanical interference. The Missouri experiment was elegant and conceptually simple, and as such it demonstrated the essence of many of these quantum experiments.

The two paths through the interferometer were arranged in the shape of a diamond. A neutron entering the diamond had its quantum wave split at the entrance, half the wave going left and half right. At the other end of the diamond, as the two waves recombined and exited, either a peak or a valley of intensity was found—just as it had on the screen of the classic Young's experiment, except that here this happened at one point rather than on a continuum of points on a screen. The scientists recorded whether they had found a peak or a trough. Then, by turning the silicon crystal, they rotated the diamond by ninety degrees so that it was vertical rather than horizontal. Now they noticed that the pattern had changed. The reason for this was that the two neutron waves were affected differently by gravity since one of them was now higher than the other, and a neutron at a higher level would

travel at a lower speed. This changed the de Broglie wavelength along one of the paths relative to the other, and hence
shifted the interference pattern. The experiment is demonstrated below.

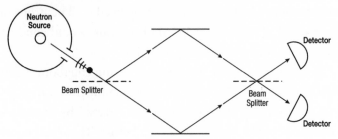

Another experiment, done by Helmut Rauch and his associates in Vienna in 1975, and also by a Missouri group that
same year, was the 2π-4π experiment with neutrons. Rauch's
Vienna team has demonstrated using neutron interferometry
a fascinating property of neutrons. A magnetic field was used
to rotate the neutron in one path of the interferometer by
360 degrees (2π). Integral-spin particles—the so-called
bosons—when undergoing a similar rotation, return to their
original state (they've thus gone around full circle); but not
so for the neutron. After turning around an angle of 360
degrees, meaning going around a full circle, the neutrons
were shown to have a sign change, which could be observed
via the interference. Only when the magnetic field rotated
the neutrons one *more* time around the circle (this is a 4π
rotation) did the neutrons return to their original state.

In Boston, Abner Shimony and Mike Horne talked during
the same period about performing this kind of experiment
with neutrons, aimed at proving the neutron's theoretically-

known 2π-4π quality—without knowing that Rauch and his students in Vienna had already performed the same experiment. Mike and Abner wrote up their paper and submitted it to a physics journal. But they soon discovered that the Vienna group had already done the same thing and had actually performed the experiment. One of Rauch's students in Vienna was Anton Zeilinger.

Anton Zeilinger was born in May 1945 in Ried/Innkreis, Austria. During the years 1963 to 1971, Anton studied physics and mathematics at the University of Vienna, receiving his Ph.D. in physics from the university in 1971, with a thesis on "Neutron Depolarization in Dysprosium Single Crystals," written under the supervision of Professor H. Rauch. In 1979, Zeilinger did his Habilitation work, on neutron and solid state physics, at the Technical University of Vienna. From 1972 to 1981, Zeilinger was a University Assistant at the Atomic Research institiute in Vienna, again working with Rauch.

Erice is a picturesque medieval town in Sicily. Physicists, no strangers to beauty and nature, have fallen in love with this small town in the stark, hilly surroundings of Sicily, and have organized annual series of conferences in this town, which attract physicists from all over the world. In 1976, the Erice conference was devoted to the foundations of quantum mechanics, including studies of Bell's inequalities and entanglement. When he got the announcement of the meeting, Rauch asked Anton Zeilinger, "Why don't you go to the meeting? We don't know much about Bell's work, but we can learn and perhaps some day perform such exciting experiments, as I hear the ones involving entanglement are, right here in Vienna . . . go and learn what you can." Anton was happy to comply and packed to go to Sicily.

At the same time, in Boston, Abner, Mike, and Frank Pipkin of Harvard were also packing their bags, ready to leave for Sicily with papers they were going to present at the meeting about their work on entanglement. Mike Horne's paper for the meeting was the one on which he and John Clauser had been working for years—an extension of Bell's theorem to probabilistic settings. In Sicily, the Boston physicists met Anton Zeilinger for the first time. "We hit it off right away," said Mike Horne. "Anton was very interested, and tried to learn from me everything he could about Bell's theorem. He was fascinated by entanglement."

One day, back in Cliff Shull's lab at the M.I.T. nuclear reactor, Cliff walked over to Mike. "Do you know a person by the name of Anton Zeilinger?" he asked, pointing to a letter in his hand. "He's just applied to come here, and mentioned your name in his letter." "Oh sure. Fantastic!" replied Mike, "He's a wonderful physicist . . . very interested in the foundations of quantum mechanics."

Anton Zeilinger joined the M.I.T. team for the 1977-78 academic year as a postdoctoral fellow, supported by a Fulbright fellowship, and over the next ten years, while he was already a Professor in Vienna, would come to Cambridge for several stints, each lasting many months. He worked hard doing the same kind of neutron diffraction work he had done as a student with Rauch in Vienna, and he and Mike Horne would co-author dozens of papers over the years, together with Cliff Shull and the students working with them at the lab at that time, the students changing from year to year. This pattern would last until Cliff Shull's retirement in 1987.

Over sandwiches, while taking breaks from the lab work, Anton and Mike would sit together discussing two-particle interference, Mike's old work with Abner and John and Dick Holt. But their current work involved performing single-neutron interference studies. The two-particle, Bell's theorem ideas were now only a passionate hobby, an interest outside their daily work. "We would sit there, having our lunch, and I would fill him in on Bell's theorem, and on local hidden variables and how they are incompatible with quantum mechanics," recalled Mike Horne, "and he would always listen and want to hear more and more."

Daniel Greenberger was born in the Bronx in 1933. He attended the Bronx High School of Science and was in the same class as Myriam Sarachik (the president-elect of the American Physical Society, now a colleague of Daniel's at CCNY), and the Nobel Prize-winning physicists Sheldon (Shelly) Glashow and Steven Weinberg. Danny subsequently studied physics at M.I.T., graduating in 1954. He then went to the University of Illinois to do doctoral work in high-energy physics with Francis Low. When Low left to take a position at M.I.T., Greenberger followed him, and wrote his dissertation at M.I.T. for his physics Ph.D. There, he studied mathematical physics, including the algebraic methods that exploit symmetries, now popular in modern theoretical physics. In the early 1960s, he joined Jeffrey Chew at the University of California at Berkeley, working on a postdoctoral fellowship in high-energy physics. He then heard that the City College of New York had opened a graduate school with a program in physics, so he moved there in 1963, and has been on the faculty there ever since.

Danny has always been fascinated by quantum theory. He maintains that quantum mechanics is not just a theory that converges with classical physics when the size of the objects in question increases. Rather, it is an independent theory with an immense richness that is not immediately apparent to us. Greenberger likens the quantum theory to the Hawaiian Islands. As we approach the islands, we only see the part that is above the water line: mountains and coastlines. But under the surface of the water there is an immense hidden dimension to these islands, stretching all the way to the bottom of the Pacific Ocean. As an example demonstrating that quantum mechanics is not an extension of classical physics but rather has this hidden dimension, Daniel Greenberger gives the idea of rotation of physical objects. Angular momentum, he reminds us, is an element of classical physics, and has an analog in quantum mechanics. But spin is something that exists only for microscopic objects that live in the quantum world and has no analog in classical physics.

Greenberger was interested in the interplay between relativity theory and quantum mechanics. In particular, he wanted to test whether Einstein's important principle of the equivalence of inertial and gravitational mass was true at the quantum level. To do so, he realized, he would need to study quantum objects that were also affected by gravity. One such object, he knew, was the neutron. Physicists have always looked for the connection between general relativity, which is the modern theory of gravity, and the quantum world.

Neutrons are quantum elements because they are small; but they are also affected by gravity. So perhaps the connection between these theories might be found by studying neutrons. Greenberger contacted the scientists working at the research reactor of the Brookhaven National Laboratory on Long Island about doing neutron research, but was told that they did not do interference studies with neutrons. He found out, however, that Cliff Shull at M.I.T. did do such research, and in 1970 Danny traveled to Cambridge to meet him. Five years later, he saw an article by Colella, Overhauser, and Werner about the Aharonov-Bohm Effect, and he contacted Overhauser and exchanged ideas with him about the effect. Danny realized there was an aspect that needed to be explored. Later he published a paper about the effect in the *Review of Modern Physics*. In 1978, there was a conference on these topics in physics at the large nuclear reactor in Grenoble, France. Overhauser, who was invited to attend the conference, couldn't go and asked Greenberger if he would go there instead.

At Grenoble, Danny met Anton Zeilinger, who at that time was working at the Grenoble reactor of the Institut Laue-Langevin as a part-time Guest Researcher. And he also met Mike Horne, who, like Danny, was attending the conference. Since Greenberger, Horne, and Zeilinger were all interested in the same topic, a bond was established among them. "That meeting changed my life," recalled Greenberger. "The three of us really hit it off together." From Grenoble, Anton went back to Austria, to continue his research there, and upon his return to M.I.T., he was pleased to find that Danny Greenberger had also joined the M.I.T. team, for a short visit. But the visit would repeat itself over and over again for many years—up until Cliff Shull's retirement in 1987—allowing the three scientists to work closely together. Even after Cliff's retirement, an N.S.F. grant together with Herb Bernstein of Hampshire College allowed them to continue their investigations.

Anton would come to M.I.T. for stays of several months, sometimes years, each; Danny would come for short visits of a few weeks at a time. The exception was Danny's long stay in 1980, when he had a sabbatical leave. The three physicists quickly became a close-knit group within the larger community of scientists doing work at the M.I.T. reactor, and they spent many hours outside the lab talking about entanglement, a topic of great interest to all of them. While at the lab they worked exclusively on one-particle (neutron) interference, many of their off-lab discussions centered on two-particle interference and Bell's magical theorem.

The entanglement among the three physicists was complete. Danny and Mike simultaneously noticed some theoretical puzzles concerning the famous Aharonov-Bohm effect of the 1950s and independently did work on the problem. Danny Greenberger wrote up his findings and published them in a journal. Anton and Danny would come up with closely related ideas about physics; and the same would happen between Mike and Anton, who for ten years would write joint papers about their research on one-particle interferometry based on their work at Shull's lab. In 1985, Mike and Anton produced a joint paper on entanglement that proposed an experiment to demonstrate that the phenomenon exists also for the positions (in addition to spin or polarization) of two particles, and that Bell's theorem would apply here too.

One day in 1985, Anton and Mike came across an announcement for a conference in Finland organized to celebrate the fiftieth anniversary of the Einstein, Podolsky, and Rosen (EPR) paper and the revolution in science it has spawned. They decided that it would be great to go to Finland, but needed a paper on two-particle interference to present at the conference; their single-particle research would not have been suitable. In a few days they had a double-diamond design for a new type of experiment to test Bell's inequality. This became their paper for the meeting. The idea was to produce entangled photons and then perform an interference experiment with these photons, using the double diamond. Their design is shown below.

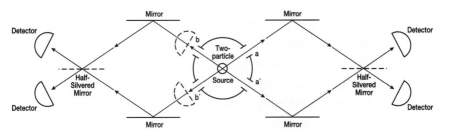

In this experimental design, a specialized source simultaneously emits two particles, A and B, traveling in opposite directions. Thus the pair can go through either holes *a* and *b*, respectively, or through holes *a'* and *b'*. Suppose that particle B is captured at one of the detectors monitoring holes *b* and *b'*. If particle B lands at *b*, then we *know* that particle A is taking hole *a*. Similarly, if particle B lands in *b'*, then we *know* that A is taking hole *a'*. Thus, for every 100 pairs produced by this source, the two upper detectors will each register 50 "A" particles; i.e., there is *no* single particle interference here, because access to particle B can reveal *which path* particle A takes. In fact, it is not even necessary to insert the detectors near holes *b* and *b'*; just the fact that we *could* determine which hole particle B takes is enough to destroy the single-particle interference for particle A.

So imagine that the detectors near *b* and *b'* are removed and the two upper ("A") and the two lower ("B") detectors are monitored while 100 pairs are emitted by the source. Quantum mechanics predicts that every detector will count 50 particles; i.e., there is no single-particle interference for either A or B because we *could* determine the route of either particle by catching the other one near the source. But quantum mechanics does predict amazing correlations between the counts. If B lands in the lower left detector, then A will certainly land in the upper right detector; if B is found in lower right, then A will land in the upper left detector. Lower left and upper left detectors *never* fire together, and neither do the lower right and upper right detectors. However, if we move one of the beam splitters an appropriate distance left or right, the correlations will change completely. The two left and the two right detectors now fire in coincidence, but detectors diagonally across from each other now never fire together. But still the count rate at each separate detector remains a steady 50, independently of the positions of the beam splitters. This behavior is explained quantum mechanically by saying that each pair of particles is emitted through *both* holes *a* and *b*, *and* through *both* holes *a'* and *b'*. This mysterious quantum state is an example of two-particle entanglement.[31]

One day, while sitting in Mike Horne's kitchen, Danny Greenberger asked him: "What do you think would happen with *three*-particle entanglement?" The question was, first of all, what are the details of three-particle correlations? The question was also: How might EPR's assumptions deal with three entangled particles? Would there be any special difficulties in trying to give a local realistic account of entanglement, or would the conflict between quantum mechanics and Einstein's locality be essentially the same as with two particles? Danny became convinced that this was a very worthwhile line of research to pursue during his upcoming sabbatical year. And, looking ahead to possible experiments, he recalled that in the Wu-Shaknov setup of positronium emission, as the two particles annihilated each other, usually two high-energy photons were emitted; but, according to the probability laws of quantum mechanics, every so often *three* photons would have to be emitted as well. This was a possible experimental setup to keep in mind during the new research project. Mike Horne thought about Danny's question, and replied, "I think that would be a great topic to pursue." Greenberger went home, and thought about the problem. Over the next few months, he would contact Mike and say: "I'm getting great results with three-particle entanglement—I have inequalities popping up everywhere; I think that three-particle entanglement may be a greater challenge to EPR than two-particle entanglement." Mike was interested, but also knew that Bell's theorem and the experiments had already proven EPR wrong, and hence there was no pressing need for another proof. But he was interested enough in the physics of three-particle entanglement to discuss the situation with Danny and encouraged him to continue.

In 1986, while Anton was back in Vienna working with Rauch, Danny was awarded a Fulbright fellowship, which allowed him to travel to Europe for his sabbatical year. He decided to use the opportunity to join Anton and work with him in Austria. The issue of three-particle entanglement remained very much on his mind as he traveled across the Atlantic. When he arrived in Vienna, Danny already had some very good ideas. He was close, he felt, to getting Bell's theorem without inequalities. In Vienna, Anton and Danny shared an office, and Danny would always show Anton his developing theoretical results, and the two of them would discuss them at length. Finally, Danny Greenberger had in front of him a situation in which a perfect correlation among three particles was enough to prove Bell's theorem. No longer was there a need to search with a partial correlation between two photons, as had been done experimentally by Clauser and Freedman, Aspect, and others. Here was a tremendously powerful—and yet conceptually simpler— proof of Bell's theorem. "Let's publish it!" said Danny, and Anton added that he and Mike had done some joint, related work that should be included in the same paper. The two conferred with Mike Horne in Boston over the phone, and decided to work on a paper on the subject.

In 1988, Mike was leafing through an issue of the journal *Physical Review Letters* at Shull's lab and noticed a paper by Leonard Mandel. The paper had an almost identical experimental design to the one he and Anton had proposed earlier in their presentation in the Finland conference. The only difference was that Mandel's two-particle interference design was a folded diamond, rather than a straight one as in the Horne-Zeilinger figure. But Mandel, who had not seen the proceedings of the Finland conference, had actually done the experiment as well; he used the down-conversion method for producing entangled photons. Thus two-particle interference was not only a thought experiment, but the real thing. And, moreover, Bell experiments could now be done with beam entanglement and without spin or polarization.

Since Anton and Mike had only presented their proposal of two-particle interference, and of Bell experiments without polarization, at conferences, and since their understanding of the entanglement basis for the interference was different and simpler than Mandel's, they decided to publish a *Physical Review Letter* presenting their results. Abner joined them in this write-up. Since the paper was essentially a comment on Mandel's breakthrough experiment, Mandel himself was assigned by the journal to be the referee. A long period of activity and cooperation followed, in which two-particle interferometry using down-conversion was pursued by the Boston team, Mandel at Rochester, Shih in Maryland, and others.

Having decided in 1986 to work together on an article on three-particle entanglement, Anton, Mike, and Danny somehow left the writing project dangling and continued their usual work. Danny Greenberger left Vienna and traveled in Europe. Eventually, his sabbatical year over, he returned to New York and to his regular teaching work. Nothing was done with the exciting new results on three-particle entanglement for two years. Then, in 1988, Danny was awarded an Alexander von Humboldt fellowship to do research in Garching, Germany, at the Max Planck Institute, where he would spend eight months as a visiting researcher. While there, he called Anton in Vienna. "Now I have the time to write . . . " he said. "I already have seventy pages," he said, "and I haven't even begun!" But the formal writing of the paper didn't proceed. Danny traveled throughout Europe, giving talks about his work with Anton and Mike on the properties of three entangled particles and how they related to Bell's theorem and EPR. At the end of the summer of 1988, Danny Greenberger went to the Erice, Sicily conference of that year. He gave a talk about three-particle entanglement, and Cornell's David Mermin — another quantum physicist — was in the audience. According to Danny, his sense was that the paper didn't really catch Mermin's attention.

But when he returned home to New York, Danny began to receive papers from several groups of physicists making references to his own work with Mike and Anton. One of these groups of physicists was headed by Michael Redhead of Cambridge University. The Redhead group claimed to have improved the Greenberger-Horne-Zeilinger work on three-particle entanglement, which Danny had presented at Erice and elsewhere in Europe. Danny called Anton and Mike: "We must do something soon," he said. "People are already referring to our work without it ever having been published."

In 1988, Danny presented a paper, which was published in the proceedings of a physics conference at George Mason

University. Meanwhile, David Mermin received the Redhead paper, which referred to the work by Greenberger, Horne, and Zeilinger. For his "Reference Frame" column in the magazine *Physics Today*, Mermin wrote an article titled "What's Wrong with These Elements of Reality?" *Physics Today* is the news magazine of the American Physical Society, and hence the paper received wide distribution. The physics community became fully aware of the new findings, referring to them as "GHZ entanglement"—even though the anticipated paper by Greenberger, Horne, and Zeilinger had still not been published. (In many sciences, a paper included in the proceedings of a conference does not count as much as a paper published in a refereed journal.) In fact, two of the paper's authors did not even know that a paper bearing their names had been presented at a conference and published in the proceedings. Danny had forgotten to mention this fact to them.

One day Abner said to Mike: "What is this thing that you and Danny and Anton proved?" "What thing?" asked Mike Horne. Abner handed him the paper by David Mermin. Mermin clearly attributed the proof he was describing, showing that quantum mechanics was incompatible with hidden variables in a strong sense in the case of three entangled particles, to Greenberger, Horne, and Zeilinger. Before he knew it, Mike was getting correspondence from the physics community congratulating him on the success of GHZ. On November 25, 1990, John Clauser wrote Mike Horne a card from Berkeley:

Dear Mike

> You old fox! Send me a (p)reprint of GHZ. Mermin seems to
> think this is super-hot stuff.

The congratulations included some from people at the top of
the profession, including Nobel Prize winners. The three
physicists quickly realized that they had better put their
research in a proper journal. To do so, they invited Abner
Shimony to join them, since he had been doing Bell work
from the beginning. In 1990, the paper, "Bell's Theorem
Without Inequalities," by Greenberger, Horne, Shimony, and
Zeilinger, was published in the *American Journal of Physics*,
although the idea of three-particle entanglement and the
improved Bell theorem continues to be called GHZ.[32]

The three-particle arrangement for presenting the GHZ
theorem can be either a spin or polarization version of the
experiment, or it can be a beam-entanglement version. The
polarization version of the GHZ experimental arrangement
is shown on page 224.

The most amazing thing about three-particle entanglement,
and the main reason for the interest taken in the GHZ proposal, is that it can be used to prove Bell's theorem without
the cumbersome use of inequalities.

The question remained: how to create *three* entangled photons in the laboratory? This can be achieved by a truly
bizarre quantum property, as was shown in a proposal by
Zeilinger and coworkers in 1997. The design is shown here.

If two pairs of entangled photons are brought into a certain experimental arrangement that makes one member of
one pair indistinguishable from one member of the other pair,

and one of the two newly-indistinguishable photons is captured, then the remaining three photons become entangled. What is so incredible here is that *the photons become entangled because an outside observer can no longer tell which pair produced the captured photon.* Then, leaving out the captured photon, the remaining three are entangled. Zeilinger and collaborators actually produced such an arrangement in 1999.

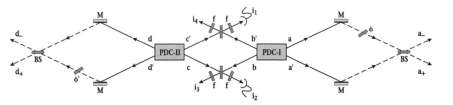

There are accessible versions of the GHZ proof of Bell's theorem using three entangled photons. David Mermin, GHZ themselves, and, recently in a textbook, Daniel Styer have presented the argument in forms suitable for a general audience.

These arguments are accessible for two common reasons. First, the quantum predictions are not derived but simply reported, thereby sparing the reader the mathematical derivations. Second, not all of the quantum predictions are reported, only the ones needed for the argument. The following version is by Mike Horne, who used it in May 2001 in his Distinguished Scholar Lecture given to the Stonehill College faculty and students. It borrows much from the earlier arguments, with the additional simplification that it uses the beam-entanglement version of GHZ, thereby avoiding spin or polarization. The argument is adapted from Mike's presentation with the kind permission and help of its author.

The figure below shows the beam-entangled GHZ setup, which clearly is a straightforward generalization of the two-particle interferometry to three particles. A half-silvered mirror at each of three locations may be set to one of two positions, the left (L) position, or the right (R) position. Depending on these settings, experimental outcomes change.

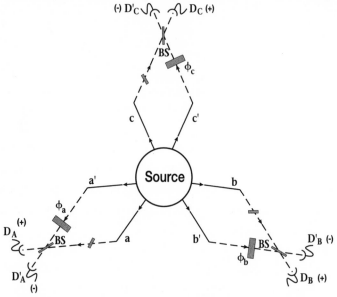

The figure shows an arrangement in which a very specialized source in the center of the figure emits three entangled particles simultaneously. Since these particles (or photons) are quantum objects, *and* they are *entangled*, each triple of particles goes *both* through holes *a*, *b*, and *c*, *and* through holes *a'*, *b'*, and *c'*. As they travel through the triple-diamond design, each particle encounters a beam-splitter (1/2-silvered mirror), which can be at either the L or the R position. Quantum mechanics predicts that, for each particle, the +1 and the -1 results (which are analogous to spin "up" or "down" for a particle, or polarization direction vertical or horizontal for a photon) occur with equal frequency: half the time +1 and half the time -1, independently of the positions of all beam splitters. If we look at pairs of particles, we still will see no interesting pattern: all pairs of results (+1, +1), (-1, -1), (+1, -1), and (-1, +1) will occur with equal frequency (1/4 of the time each) for both particles A and B (and similarly for the other pairs, B and C and A and C), independently of the positions of the beam splitters. However, quantum mechanics predicts that an observer will see a truly magical dance if the observer should look at what happens to *all three particles*. For example, quantum mechanics predicts that if the beam splitters for particles B and C are both set in the L position, and both of these particles land in, say, the -1 outcome detectors, and if the particle A beam splitter is set to the R position, then particle A will land in the +1 detector with certainty. This is a remarkably strong prediction, and there are similar perfect predictions for other combinations of settings. The table below summarizes the combinations of settings and the quantum mechanical predictions.

For beam splitter settings:			*The quantum mechanical predictions are:*
	A	B	C	

	A	B	C	
1.	R	L	L	Either 0 or 2 particles will go to -1
2.	L	R	L	Either 0 or 2 particles will go to -1
3.	L	L	R	Either 0 or 2 particles will go to -1
4.	R	R	R	Either 1 or 3 particles will go to -1

Other setting combinations (for example, LLL) are not needed in our discussion.

The predictions on the right for the particular setting combinations on the left were obtained by Greenberger, Horne, Shimony, and Zeilinger using the mathematics of quantum mechanics. They began, of course, with the actual entanglement state of the three particles. The idea of entanglement is a superposition of states, as we know, and for three particles, each going through *two* apertures, we have the superposition state that can be written (in a somewhat simplified form) as:

$$(abc + a'b'c')$$

This equation is the mathematical statement of three-particle entanglement, in which the "+" sign captures the *both-and* property mentioned earlier.

From the equation, which describes the superposition of the states—i.e., describes mathematically exactly what it means for three particles to be entangled, within the specific setting of this experiment with its six holes—the physicists worked out the mathematics and derived the predictions, listed in the table above. The actual details can be found in the appendix to the paper "Bell's Theorem Without Inequalities," by Greenberger, Horne, Shimony, and Zeilinger, *American Journal of Physics*, 58 (12), December 1990. Note that even in their scientific paper, the authors relegated their algebraic derivation of the quantum mechanical predictions based on the state equation to an appendix—it was just too long, and it is elementary quantum mechanics. The interested (and mathematically inclined) reader may look for these details there. What is important for the reader to understand is that the predictions in the table above are exactly what quantum mechanics tells us will happen in each situation. There is nothing more in these predictions than an application of the rules of quantum mechanics to a particular setting and the state of entanglement of the three particles. We will therefore take these predictions as valid, direct consequences of the entanglement of the three particles.

Going back to our table of the quantum mechanical predictions for the three-particle entangled state, we find that: *Given the beam-splitter settings, and given specific outcomes for B and C, the outcome for particle A is predictable with certainty.* For example, suppose that the beam-splitters for particles B and C are both in the L position, and that particle B lands in the -1 detector and C also lands in -1. Then, if the particle A beam-splitter is in the R position, particle A will certainly go to detector +1. There are similar perfect correlations, as can be seen from the table above, for other choices of beam-splitter settings and other outcomes at two stations. *In short, given the beam-splitter settings and specific outcomes for B and C, the outcome for particle A is predictable with certainty.*

Now comes the important part of the work of GHZ. To understand what it is, and why the GHZ state provides such a powerful demonstration and extension of Bell's theorem, we have to go back to what Einstein and his colleagues said fifty-five years earlier, in the EPR paper of 1935.

Einstein and his coworkers noted the strikingly perfect correlations present in a theoretical two-particle entanglement. They argued that these perfect correlations are perplexing— unless they simply reveal pre-existing, objectively-real properties of the entangled objects. Einstein and his colleagues stated their commitment to the existence of an objective reality as follows (in the EPR paper of 1935):

"If, without in any way disturbing a system, we can predict with certainty the value of a physical quantity, then there exists an element of physical reality corresponding to this physical quantity."

Now, the landing of particle A in its +1 detector is an "element of reality" as per Einstein's definition, because we can *predict that this will happen with certainty, and clearly we did not disturb particle A by our choice of beam splitter settings at the distant locations B and C.* The outcome at A can at most depend on the beam splitter setting at station A, not at B or C. Now, since the landing of particle A in detector +1 is an "element of reality," let's call this element of reality A(R). Thus, A(R) is the element of reality at location A. It signifies the outcome at station A when the beam splitter that controls particle A is set to the right (R) setting. For the specific outcome that particle A lands in the +1 detector, we say that the element of reality is +1 and write it as: A(R)=+1. Similarly for other locations and settings combinations, we have, following Einstein, six elements of reality: A(R), B(R), C(R), A(L), B(L), and C(L). Each of these elements of reality has a value of either +1 or -1.

Now comes the GHZ Theorem:

Assume that Einstein's elements of reality *do exist* and can explain the otherwise baffling quantum mechanical predictions given in the table above (and which, by now, have been experimentally verified by an actual 3-particle entanglement experiment conducted by Zeilinger in 1999). Agreement with quantum predictions 1, 2, 3, and 4 of the table above imposes the following constraints on the elements of reality:

1. A(R) B(L) C(L)= +1

2. A(L) B(R) C(L)= +1

3. A(L) B(L) C(R)= +1

4. A(R) B(R) C(R)= -1

The above statements are true because of the following. In case (1), the settings are RLL and, according to quantum mechanics, as listed in the table earlier, "Either 0 or 2 particles go to -1." Thus either 0 or two of the elements of reality A(R), B(L), and C(L) are equal to -1. And when you multiply all three of them, you will thus get: 1x1x1=1 (in the case 0 of them go to -1) or 1x(-1)x(-1)=1 (for the case that 2 particles go to -1; regardless of order). Similarly, for cases (2) and (3) we also get that the product of the elements of reality is equal to 1, either because all three of them are equal to 1 (0 particles go to -1), or because any two of them are -1 (the case that 2 particles go to -1) and the third is a +1.

In case (4), the quantum mechanical prediction is that either 1 particle or 3 particles go to -1. Thus the possible products of the three elements of reality A(R) B(R) and C(R) are:

-1 times two +1s, or three -1s multiplied together. In either case, the product has an odd number of -1s and the answer therefore is -1.

Now comes the great trick: *Multiply together the top three equations.* Multiplication of the left sides gives us:

$$A(R) \ A(L) \ A(L) \ B(L) \ B(R) \ B(L) \ C(L) \ C(L) \ C(R) = A(R) \ B(R) \ C(R)$$

The reason that this is true is that each of the terms excluded from the right side of the equation appears *twice* on the left side of the equation. Each of the terms A(L), B(L), C(L) has a value of either +1 or -1; when such a term appears twice in the equation, the product of the term times itself is certainly equal to +1 (because +1x+1=+1 and -1x-1=+1).

Now, multiplying the right sides of the equations (1), (2), and (3) we get +1x+1x+1=+1; so that we have: A(R) B(R) C(R) = +1.

But our quantum mechanical prediction, equation (4), says that: A(R) B(R) C(R) = -1.

Thus we have a contradiction. Therefore, Einstein's "elements of reality" and locality could not possibly exist if quantum mechanics is correct. Hidden variables are impossible within the framework of quantum mechanics. The entangled particles do not act the way they do because they were "pre-programmed" in any way: such programming is impossible if particles behave according the rules of the quantum theory. The theorem shows that any instruction sets the particles might possess must be internally inconsistent, and hence impossible. The particles respond *instantaneously* across any distance separating them in order to give us the results that quantum theory says will be obtained. This is the magic of entanglement.

Furthermore, actual experiments have shown that the quantum theory is correct, and therefore Einstein's local realism is not. The GHZ theorem proves the contradiction in a much more direct, easier to understand, and non-statistical way, as compared with Bell's original theorem.

"In all our work, there has never been any competition. It's been wonderful," recalled Mike Horne when describing to me his work with his colleagues in coming up with the GHZ design and the discovery of the GHZ triple-particle entangled state. "We were fortunate to work in a field in which very few people were working, and thus everyone welcomed others who were excited about the same problems in the foundations of quantum mechanics," he said.

These physicists, working together in harmony, produced one of the most important contributions to modern physics. Their work would be expanded and extended in the following years, and it would help spawn new technologies, which could only have been imagined by science fiction writers only a few years earlier.

The Borromean rings are named after the Borromeo family, whose members belong to the Italian nobility. The family owns the beautiful Borromean Islands on Lake Maggiore in northern Italy. The family coat of arms consists of three rings intertwined in an interesting way: should one of them be broken, the other two will no longer remain linked as well. The rings may represent the idea of "united we stand, divided we fall." The physicist P.K. Aravind has studied entanglement and has discovered connections between entangled stated in quantum mechanics and various kinds of topological knots. In particular, Aravind has argued that there is a one-to-one correspondence between the GHZ entangled state of three particles and the Borromean rings. The Borromean rings are shown below.[33]

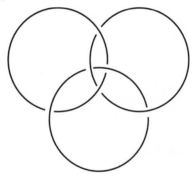

Aravind's proof has to do with entanglement along a particular direction of spin (the z-direction). He's also shown that if one measures the spin of three entangled particles along another direction, the x-direction, then the entangled state is different. Now it is no longer analogous with the Borromean rings, but rather with the Hopf rings. Three Hopf rings are interlocked in such a way that if one of them is cut, the other two remain locked together. Three Hopf rings are shown below.

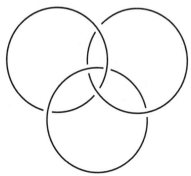

Aravind has also demonstrated that a general, n-particle GHZ state of entanglement could be viewed as a generalization of the three Borromean rings. Such a linking of several particles is analogous to a linked chain that looks like the rings below.

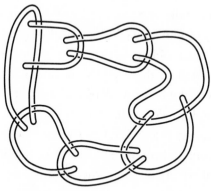

Danny Greenberger still spends time alternately visiting Mike Horne in Boston and Anton Zeilinger in Vienna, thus

18

The Ten-Kilometer Experiment

"If two separated bodies, each by itself known maximally, enter a situation in which they influence each other, and separate again, then there occurs regularly that which I have just called *entanglement* of our knowledge of the two bodies."

—Erwin Schrödinger

T
he next chapter in the history of the mysterious phenomenon of entanglement was written by Nicholas Gisin of the University of Geneva. Gisin was born in Geneva in 1952 and studied theoretical physics at the University of Geneva, obtaining his Ph.D. in this field. He was always interested in the mystery of entanglement. In the 1970s, he met John Bell at CERN, and was very much taken with the man, later describing him as sharp and impressive. Gisin immediately recognized Bell's work as a groundbreaking achievement in theoretical physics. Gisin wrote a number of theoretical papers on Bell's theorem, proving important results about quantum states. He then spent some time at the University of Rochester, where he met some of the pioneers in optics research: Leonard Mandel, whose work made him

keeping alive the entanglement among these three good friends. In Austria, Danny spends time with Anton's research group at the University of Vienna—a key group conducting leading-edge work on a wide array of quantum behavior and entanglement, including teleportation. Recently, Danny attended a party given by the research group. There he met Schrödinger's daughter, and, by her side, Schrödinger's grandson—by another mother. The young man, a member of the research group, had not found out that the great physicist was his grandfather until he became an adult and a quantum physicist himself.

a legend in the field, and Emil Wolf.

Nicholas then returned to Geneva and worked in industry for four years. This was a fortuitous move since it allowed him to combine his passion for quantum mechanics with practical work with fiber optics. The link he forged between fiber optics technology and quantum theory would prove crucial for the new work on entanglement. Equally important would be the connections he established with telephone companies. Returning to the University of Geneva, Gisin began to design experiments to test Bell's inequality.

By that time, the 1990s, Clauser and Friedman and others had established the first experimental violation of Bell's inequality and Alain Aspect had taken the work further than anyone by establishing that any signal from one point of the experimental setup to the other would have had to travel at a speed faster than light, thus establishing that no such signal could have been received. Aspect's experiment was done within the space of a laboratory. Following Aspect's experiments, Anton Zeilinger and collaborators have extended the range at which entanglement was tested to hundreds of meters, across several buildings around their laboratory in Austria. This setup is shown below.

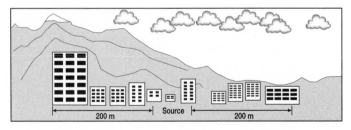

But Gisin wanted to go much farther. First, he designed an experiment by which entangled photons traveled a distance of 35 meters, inside his laboratory.

His connections with the telephone companies allowed him to enlist their enthusiastic support for an ambitious experiment. The scale of the work would be unprecedented: Gisin conducted his photon experiment not in air but within a fiber-optical cable. And the cable was laid from one location to another, 10.9 kilometers (seven miles) away as the crow flies. Counting the actual distance traveled, with all the bends and curvature of the cable, one reaches a total distance of 16 kilometers (ten miles). Gisin came to the experiment with an open mind. He would have found either outcome fascinating: a confirmation of quantum mechanics or a result supporting Einstein and his colleagues. The result was an overwhelming affirmation of entanglement, the "spooky action at a distance," which Einstein so disliked. Bell's inequality was once again used to provide strong support for nonlocality. Because of the experimental setup, a signal from one end of the cable to the other, telling one photon what setting the other photon found, would have had to travel at *ten million times the speed of light*. The map of the experiment is shown below.

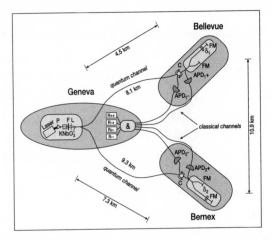

Like some other physicists, Gisin believes that while entanglement doesn't allow us to send readable messages faster than light, the phenomenon still violates the *spirit* of special relativity. He thus wanted to test the entanglement phenomenon within a relativistic framework. In one of his experiments, Gisin used an absorbing black surface, placed at the ends of the optical fiber, to collapse the wave function. The two ends of the fiber through which entangled photons were to appear were again placed kilometers apart, but the absorbing surfaces were moved at extremely high speeds. By manipulating these experimental conditions, it was possible to study the entanglement phenomenon using different relativistic reference frames. Thus time itself could be manipulated in accordance with the special theory of relativity: each photon could be measured as arriving at its endpoint at different times. First, one member of a pair of photons was the first one to arrive at its target, and in the second experiment its twin arrived before it. This complex experiment using moving reference frames resulted in a strong confirmation of nonlocal entanglement and the predictions of quantum mechanics.

In the 1990s, the big news in quantum technology was cryptography. The idea of using entanglement in quantum cryptography was put forward by Arthur Ekert of Oxford University in 1991. The term is a bit of a misnomer since cryptography is the art of encrypting messages. Quantum cryptography, however, usually means techniques for evading and detecting eavesdroppers. Entanglement plays an important role within this new technology. Gisin's associates at the Swiss telephone companies were very interested in this kind of research, since it could allow for the development of secure communications networks. He performed research in quantum cryptography, and in one of his recent experiments was able to transmit secure messages a distance of 25 kilometers (16 miles) under the water of Lake Geneva. Gisin is enthusiastic about his great achievements in cryptography, both using entanglement and using other methods. He believes that the field has matured and that quantum cryptography could be used commercially at distances such as the ones used in his experiments. Gisin has also spent time in Los Alamos, where an American team of scientists is making progress on quantum computing, another proposed new technology that—if successful—would use entangled entities.

19

Teleportation: "Beam me up, Scotty!"

"Entanglement—along with superposition of states—is the strangest thing about quantum mechanics."
—William D. Phillips

Quantum teleportation has until recently been only a thought experiment, an idea that had never been successfully tested in the real world. But in 1997, two teams of scientists were successful in realizing the dream of teleporting a single particle's quantum state.

Quantum teleportation is a way of transferring the state of one particle to a second particle, which may be far away, effectively *teleporting* the initial particle to another location. In principle, this is the same idea—at this point existing only within the realm of science fiction—by which Captain Kirk can be teleported back into the spaceship *Enterprise* by Scotty, who is aboard the spaceship.

Teleportation is the most dramatic application we can imagine of the phenomenon of entanglement. Recently, two international teams, one headed by Anton Zeilinger in Vienna, and the other headed by Francesco De Martini in Rome, brought the idea of teleportation from the imagination to reality. They followed a suggestion made in 1993 by Charles Bennett in an article in a physics journal. Bennett showed that there was a physical possibility of teleporting the quantum state of a particle.

The reason physicists began to think about teleportation was that in the 1980s it was shown by William Wootters and W. Zurek that a quantum particle can never be "cloned." The *No Cloning Theorem* of Wootters and Zurek says that if we have a particle, its state cannot be *copied* onto another particle, *while the original particle remains the same*. Thus, it is impossible to create a kind of copying machine that would take one particle and imprint its information onto another particle, keeping the original intact. Thus the only way that physicists could conceive of imprinting information from one particle onto another was by having the same information disappear from the original particle. This hypothetical process was later given the name teleportation.

The paper describing the dramatic teleportation experiment of Zeilinger's team, "Experimental quantum teleportation," by D. Boumeester, J.-W. Pan, K. Mattle, M. Eibl, H. Weinfurter, and A. Zeilinger, appeared in the prestigious journal *Nature* in December 1997. It says:

"The dream of teleportation is to be able to travel by simply reappearing at some distant location. An object to be teleported can be fully characterized by its properties, which in classical physics can be determined by measurement. To make a copy of that object at a distant location one does not need the original parts and pieces—all that is needed is to send the scanned information so that it can be used for reconstructing the object. But how precisely can this be a true copy of the original? What if these parts and pieces are electrons, atoms and molecules?" The authors discuss the fact that since these microscopic elements making up any large body are given to the laws of quantum mechanics, the Heisenberg uncertainty principle dictates that they cannot be measured with arbitrary precision. Bennett, et al., suggested the idea of teleportation in an article in *Physical Review Letters* in 1993, proposing that it may be possible to transfer the quantum state of a particle to another particle—a *quantum* teleportation—provided that the person doing the teleportation *does not obtain any information about the state in the process.*

It seems absurd that any information obtained by an outside observer should affect what goes on with a particle, but according to quantum mechanics, the mere process of *observing* a particle destroys (or "collapses") the wave-function of the particle. Properties of momentum and position, for example, cannot be known to any given precision. Once measured (or otherwise actualized), a quantum object is no longer in that fuzzy state in which quantum systems are, and information is thus destroyed in the process of being obtained.

But Bennett and his coworkers had a brilliant idea as to how one might transfer the information in a quantum object without measuring it, i.e., without collapsing its wave-function. The idea was to use entanglement. Here is how teleportation works.

Alice has a particle whose quantum state, unknown to her, is Q. Alice wants Bob, who is at a distant location, to have a particle in the same state as her particle. That is, Alice wants Bob to have a particle whose state will also be Q. If Alice measures her particle, this would not be sufficient since Q cannot be fully determined by measurement. One reason is the uncertainty principle, and another is that quantum particles are in a superposition of several states at the same time. Once a measurement is taken, the particle is forced into one of the states in the superposition. This is called the projection postulate: the particle is projected onto one of the states in the superposition. The projection postulate of quantum mechanics makes it impossible for Alice to measure the state, Q, of her particle in such a way that she would obtain *all* the information in Q, which is what Bob would need from her in order to reconstruct the state of her particle on his own particle. As usual in quantum mechanics, observing a particle destroys some of its information content.

This difficulty, however, can be overcome by a clever manipulation, as Bennett and his colleagues understood. They realized that precisely the projection postulate enables Alice to teleport her particle's state, Q, to Bob. The act of teleportation sends Bob the state of Alice's particle, Q, while destroying the quantum state for the particle she possesses. This process is achieved by using a pair of entangled particles, one possessed by Alice (and it is not her original particle with state Q), and the other by Bob.

Bennett and his colleagues showed that the full information needed so that the state of an object could be reconstructed is divided into two parts: a quantum part and a classical part. The quantum information can be transmitted instantaneously—using entanglement. But that information cannot be used without the classical part of the information, which must be sent through a classical channel, limited by the speed of light.

There are, therefore, two channels for the teleportation act: a quantum channel and a classical channel. The quantum channel consists of a pair of entangled particles: one held by Alice and the other held by Bob. The entanglement is an invisible connection between Alice and Bob. The connection is delicate, and must be preserved by keeping the particles isolated from their environment. A third party, Charlie, gives Alice another particle. The state of this new particle is the message to be sent from Alice to Bob. Alice can't read the information and send it to Bob, because—by the rules of quantum mechanics—the act of reading (measurement) alters the information unpredictably, and not all the information can be obtained. Alice measures a joint property of the particle Charlie has given her and her particle entangled with Bob's. Because of this entanglement, Bob's particle responds immediately, giving him this information—the rest of it Alice communicates to Bob by measuring the particle and sending him that partial information through a classical channel. This information tells Bob what he needs to do with his entangled particle in order to obtain a perfect transformation of Charlie's particle into his own, completing the teleportation of Charlie's particle. It is noteworthy that neither Alice nor Bob ever know the state that one has sent and the other received, only that the state has been transmitted. The process is demonstrated in the figure below.

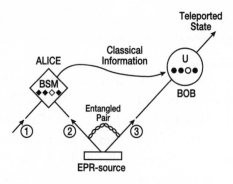

Can teleportation be extended to larger objects, such as people? Physicists are generally reluctant to answer such a question, viewing it as beyond the scope of physics today, and perhaps in the realm of science fiction. But many scientific and technological developments have been considered fantasy until they became a reality. Entanglement itself was thought to be within the realm of the imagination until science proved that it is a real phenomenon, despite its bizarre nature.

If teleportation of people or other large objects should be possible, can we envision how this might be done? This question, and the previous one, touch upon one of the greatest unsolved problems in physics: Where does the boundary lie separating the macro-world we know from everyday life and the micro-world of photons, electrons, protons, atoms, and molecules?

We know from de Broglie's work that particles have a wave-aspect to them, and that the wavelength associated with a particle can be computed. Thus, in principle, even a person can have an associated wave-function. (There is another technical point here, which is beyond what we can discuss in this book, and it is that a person or another macroscopic object would not be in a pure state, but rather in a "mixture" of states). The answer to the question as to how the teleportation of a person might be carried out can be restated as the question: Is a person the sum of many elementary particles, each with its own wave-function, or a single macro-object with a single wave-function (of a very short wavelength)? At this point in time, no one has a clear answer to this question, and teleportation is therefore still a real phenomenon only within the realm of the very small.

Chapter 20

Quantum Magic:
What Does It All Mean?

"The conclusions from Bell's theorem are philosophically startling; either one must totally abandon the realistic philosophy of most working scientists or dramatically revise our concept of space-time."

—Abner Shimony and John Clauser

"So farewell, elements of reality!"

—David Mermin

What does entanglement mean? What does it tell us about the world and about the nature of space and time? These are probably the hardest questions to answer in all of physics.

Entanglement breaks down all our conceptions about the world developed through our usual sensory experience. These notions of reality are so entrenched in our psyche that even the greatest physicist of the twentieth century, Albert Einstein, was fooled by these everyday notions into believing that quantum mechanics was "incomplete" because it did not include elements he was sure had to be real. Einstein felt that what happens in one place could not possibly be directly and instantaneously linked with what happens at a distant location. To understand, or even simply accept, the validity of entanglement and other associated quantum phenomena, we must first admit that our conceptions of reality in the universe are inadequate.

Entanglement teaches us that our everyday experience does not equip us with the ability to understand what goes on at the micro-scale, which we do not experience directly. Greenstein and Zajonc (*The Quantum Challenge*) give an example demonstrating this idea. A baseball hit against a wall with two windows cannot get out of the room by going through *both* windows at once. This is something every child knows instinctively. And yet an electron, a neutron, or even an atom, when faced with a barrier with two slits in it, will go through both of them at once. Notions of causality and of the impossibility of being at several locations at the same time are shattered by the quantum theory. The idea of superposition—of "being at two places at once"—is related to the phenomenon of entanglement. But entanglement is even more dramatic, for it breaks down our notion that there is a meaning to spatial separation. Entanglement can be described as a superposition principle involving two or more particles. Entanglement is a superposition of the states of two or more particles, taken as one system. Spatial separation as we know it seems to evaporate with respect to such a system. Two particles that can be miles, or light years, apart may behave in a concerted way: what happens to one of them happens to the other one instantaneously, regardless of the distance between them.

WHY CAN'T WE USE ENTANGLEMENT TO SEND A MESSAGE FASTER THAN LIGHT?

Entanglement may violate the spirit of relativity, but not in a way that allows us to use it to send a message faster than light. This is a very important distinction, and it captures in its core the very nature of quantum phenomena. The quantum world is random in its nature. When we measure, we force some quantum system to "choose" an actual value, thus leaping out of the quantum fuzz into a specific point. Thus, when Alice measures the spin of her particle along a direction she chooses (or, equivalently, measures the polarization of a photon along a direction she chooses), she *cannot choose the result.* The result will be "up" or "down," but Alice cannot predict what it will be. Once Alice makes the measurement, Bob's particle or photon is forced into a particular state (opposite spin along that direction, for a particle; same polarization direction for a photon). But since Alice has no control over the result she gets, she can't "send" any meaningful information to Bob. All that can happen because of the entanglement is as follows. Alice can choose any one of many possible measurements to carry out, and, whichever one she chooses, she will get a result. But she doesn't know ahead of time which of two results she will get. Similarly, Bob can choose any one of many measurements to make and doesn't know the result ahead of time. But, because of the entanglement, if they happen to have chosen the same measurement, their unpredictable results will be opposite (assuming a spin measurement).

Only after comparing their results (using a conventional method of communication, which cannot send information faster than light) can Alice and Bob see the coincidence of their results.

On the face of it, there is nothing problematic about strong correlations; one simply introduces "elements of reality" to explain them, as Einstein wanted to do. But Bell's proof leads us to the conclusion that this approach doesn't work.

Abner Shimony has referred to entanglement as "passion at a distance," in an effort to avoid the trap of assuming that one can somehow use entanglement to send a message faster than light. Shimony believes that entanglement still allows for quantum mechanics and relativity theory to enjoy a "peaceful coexistence," in the sense that entanglement does not violate special relativity in a strict sense (no messages can travel faster than light). Other physicists, however, believe that the "spirit of relativity theory" still is violated by entanglement, because "something" (whatever it may be) does "travel" faster than light (in fact, infinitely fast) between two entangled particles. The late John Bell was of this belief.

Possibly a way to understand entanglement is to avoid looking at relativity theory altogether, and not to think of two entangled entities as particles "sending a message" from one to the other. In a paper entitled "Quantum Entanglement," Yanhua Shih argues that because two entangled particles are (in some sense) not separate entities, there is even no apparent violation of the uncertainty principle, as EPR had suggested.

Entangled particles transcend space. The two or three entangled entities are really parts of one system, and that system is unaffected by physical distance between its components. The system acts as a single entity.

What is fascinating about the quest for entanglement is that a property of a quantum system was first detected by mathematical considerations. It is amazing that such a bizarre, other-worldly property would be found mathematically, and it strengthens our belief in the transcendent power of mathematics. After the mathematical discovery of entanglement, clever physicists used ingenious methods and arrangements to verify that this stunning phenomenon does actually occur. But to truly understand what entanglement is and how it works is for now beyond the reach of science. For to understand entanglement, we creatures of reality depend on "elements of reality," as Einstein demanded, but as Bell and the experiments have taught us, these elements of reality simply do not exist. The alternative to these elements of reality is *quantum mechanics*. But the quantum theory does not tell us *why* things happen the way they do: *why* are the particles entangled? So a true comprehension of entanglement will only come to us when we can answer John Archibald Wheeler's question: "Why the quantum?"

Acknowledgments

I am most grateful to Abner Shimony, Professor Emeritus of Physics and Philosophy at Boston University, for his many hours of help, encouragement, and support to me while I was preparing this work. Abner has unselfishly allowed me to borrow many papers, books, conference proceedings, as well as letters and manuscripts from his personal collection relevant to quantum theory and entanglement. Abner spared no effort in answering my myriad questions on entanglement and the magic of quantum mechanics, explaining to me many obscure mathematical and physical facts about the mysterious quantum world, and telling me the story of his own role in the quest for entanglement, as well as many anecdotes about the search for an understanding of this amazing phenomenon. Abner and I talked for many hours at his home, in the car, over coffee at a restaurant, while taking a walk together, or late at night on the telephone and I greatly appreciate his labor of love in helping me get this story straight, as well as going over the manuscript and offering many suggestions for improvement and clarification.

I wish to express my deep appreciation to Michael Horne, Professor of Physics at Stonehill College in Massachusetts, for sharing with me the details of his work with Abner Shimony on designing an experiment to test Bell's inequality, his important work on one- two- and three-particle interferometry, and his groundbreaking work on three-particle entanglement with Daniel Greenberger and Anton Zeilinger, widely known as the GHZ design. I am extremely grateful to Mike for his many hours of help to me while I was preparing this manuscript, for answering my many questions, and for providing me access to many important papers and results. Mike carefully went over the manuscript, corrected many of my errors and inaccuracies, and offered many suggestions for improvement. I also thank him for kindly allowing me to adapt material on three-particle entanglement from his Distinguished Scholar Lecture at Stonehill College for use in this book. Thank you, Mike!

I am very grateful to Alain Aspect of the Center for Optics Research at the University of Paris in Orsay for explaining his important work to me, and for teaching me some fine points of the theory of entangled states. Alain most kindly opened up his laboratory to me, showing me how he designed his historic experiments, built his own complex devices, and how he obtained his stunning results on entangled photons. I thank Professor Aspect for his time and effort and excitement about physics. Merci, AA.

John Clauser and his colleague Stuart Freedman actually carried out the first experiment designed to test Bell's theory at Berkeley in 1972, based on the joint work with Mike Horne, Abner Shimony, and Richard Holt (the famous CHSH paper). I thank John Clauser for sharing with me the results of his experiments and for providing me with many important papers on the topic of entanglement and for several thought-provoking interviews.

In the years following the Clauser and Aspect experiments, a number of physicists around the world derived further results demonstrating the existence of entangled particles and light waves. Nicholas Gisin of the University of Geneva produced entangled photons at great distance. Gisin has demonstrated entangled states for photons that were 10 kilometers apart, as well as studied numerous properties of entangled states and their use in quantum cryptography and other applied areas. He is also known for important theoretical work on Bell's theorem. Nicholas Gisin generously shared with me the results of his experiments, and I thank him warmly for providing me with many research papers produced by his group at the University of Geneva, as well as for informative interviews.

The implications of entanglement are far-reaching, and scientists are currently exploring its implications in quantum computing and teleportation. Anton Zeilinger of the University of Vienna is a leading scientist in this area. He and his colleagues have demonstrated that teleportation is possible, at least for photons. Anton Zeilinger's work spans several decades and includes the pioneering work on three-particle entanglement, work done jointly with Greenberger and

Horne (GHZ), as well as entanglement swapping and other projects demonstrating the strange world of tiny particles. I am very grateful to Anton for providing me with much information on his work and achievements. Also in Vienna, I am grateful to Ms. Andrea Aglibut of Zeilinger's research group for providing me with many papers and documents relevant to the work of the group.

I am grateful to Professor John Archibald Wheeler of Princeton University for welcoming me at his house in Maine and for discussing with me many important aspects of quantum theory. Professor Wheeler generously shared his thoughts about quantum mechanics and its role in our understanding of the workings of the universe. His vision of quantum mechanics in the wider context of physics and cosmology shed much important light on the questions raised by Einstein, Bohr, and others about the meaning of physics and its place in human investigations of nature.

I am grateful to Professor Yanhua Shih of the University of Maryland for an interesting interview about his many research projects relevant to entanglement, teleportation, and the method of parametric down conversion. Professor Shih and his colleagues were instrumental in producing some of the most stunning evidence for the effects of entanglement. I thank Yanhua for sharing with me his many research papers.

I thank Professor Daniel Greenberger of the City University of New York for information on the amazing GHZ experimental design and the theoretical demonstration he has provided jointly with Horne and Zeilinger of Bell's theorem in a simple and dramatic way. I am grateful to Danny for much information about his work.

I am grateful to Professor William Wootters of Williams College in Massachusetts for an interesting interview about his work and on his joint "no-cloning theorem." The Wootters-Zurek theorem, which proves that there can be no quantum-mechanical "copy machine" that preserves the originals, has important implications in quantum theory, including teleportation.

I thank Professor Emil Wolf of the University of Rochester for a discussion of the mysteries of light, and for important details about his work and the work of his late colleague Leonard Mandel, whose pioneering achievements exposed many puzzling properties of entangled photons.

I am very grateful to Professor P.K. Aravind of the Worcester Polytechnic Institute in Massachusetts for sharing with me his work on entanglement. Professor Aravind has demonstrated surprising consequences of Bell's theorem and entangled states in a number of theoretical papers. Thank you, P. K., for sharing your work with me and for explaining to me some aspects of quantum theory.

I thank Professor Herbert Bernstein of Hampshire College in Massachusetts for an interesting interview about the meaning of entanglement. I am also grateful to Herb for pointing out to me the German origin and meaning of the original term used by Erwin Schrödinger to describe the phenomenon.

I am grateful to Dr. William D. Phillips of the National Institute of Standards and Technology, a Nobel Prize winner in physics, for an interesting discussion of the mysteries of quantum mechanics and the phenomenon of entanglement, as well as for providing me with interesting details of his work in quantum mechanics.

Dr. Claude Cohen-Tannoudji met with me in Paris and was both most generous with his time and informative. Cohen-Tannoudji is co-author of one of the classic textbooks in the field, *Quantum Mechanics,* a work that he and his colleagues spent five years finetuning. I thank him for his kindness, in readily sharing his expertise with me.

I thank the physicist Dr. Mary Bell, John Bell's widow, for cooperation in my preparation of the material relating to the life and work of her late husband.

I am grateful to Ms. Felicity Pors of the Niels Bohr Institute in Copenhagen for help in facilitating the use of historical photographs of Niels Bohr and other physicists.

None of the experts acknowledged above are to be blamed for any errors or obscurities that may have remained in this book.

I thank my publisher and friend, John Oakes, for his encouragement and support during the time I was writing this book. I thank the dedicated staff at Four Walls Eight Windows: Kathryn Belden, Jofie Ferrari-Adler, and John Bae, for their help and dedication in producing this book. I thank my wife, Debra, for her help and encouragement.

Notes

1. Note, however, that causation is a subtle and complicated concept in quantum mechanics.

2. *The New York Times,* May 2, 2000, p. F1.

3. Richard Feynman. *The Feynman Lectures.*Vol. III. Reading, MA: Addison-Wesley, 1963.

4. As reported by Abraham Pais. *Niels Bohr's Times.* Oxford: Clarendon Press, 1991.

5. Much of the biographical material in this chapter is gleaned from Walter Moore. *Schrödinger: Life and Thought.* New York: Cambridge University Press, 1989.

6. Walter Moore. *Schrödinger: Life and Thought.* New York: Cambridge University Press, 1989.

7. M. Horne, A. Shimony, and A. Zeilinger, "Down-conversion Photon Pairs: A New Chapter In the History of Quantum Mechanical Entanglement," *Quantum Coherence,* J.S. Anandan, ed., Singapore: World Scientific, 1989.

8. E. Schrödinger, *Collected Papers on Wave Mechanics*, New York: Chelsea, 1978, 130.

9. E. Schrödinger, *Proceedings of the Cambridge Philosophical Society*, 31 (1935) 555.

10. Armin Hermann. *Werner Heisenberg 1901-1976*. Bonn: Inter Nations, 1976.

11. Author's interview with John Archibald Wheeler, June 24, 2001.

12. Wheeler, J. A., "Law without Law," contained in the collection of papers, *Quantum Theory and Measurement*, edited by J.A. Wheeler and W. H. Zurek. Princeton, NJ: Princeton University Press, 1983.

13. John Archibald Wheeler, "Law Without Law," Wheeler and Zurek, eds., p. 182-3.

14. John Archibald Wheeler, "Law Without Law," Wheeler and Zurek, eds., p. 189.

15. Much of the biographical information in this chapter is adapted from Macrae, Norman. *John Von Neumann: The Scientific Genius Who Pioneered the Modern Computer, Game Theory, Nuclear Deterrence, and Much More*. Providence, R.I.: American Mathematical Society, 1992.

16. See Amir D. Aczel. *God's Equation*. New York: Four Walls Eight Windows, 1999.

17. A. Fölsing, *Albert Einstein*, New York: Viking, 1997, p. 477.

18. Louis de Broglie. *New Perspectives in Physics*. New York: Basic Books, 1962, p. 150.

19. Reported in J. A. Wheeler and W. H. Zurek, eds. *Quantum Theory and Measurement*. Princeton, NJ: Princeton University Press, 1983, p. viii.

20. Quoted in Wheeler and Zurek, 1983, p.7.

21. Abraham Pais. *Niels Bohr's Times*. New York: Clarendon Press, 1991, p.427.

22. Wheeler and Zurek, p. 137.

23. Albert Einstein, Boris Podolsky, and Nathan Rosen, "Can Quantum-Mechanical Description of Physical Reality Be Considered Com-

plete?" *Physical Review*, 47, 777-80 (1935).

24. Pais, 1991, p. 430.

25. Wheeler and Zurek, 1983.

26. A. Einstein, B. Podolsky, and N. Rosen, "Can Quantum-Mechanical Description of Physical Reality Be Considered Complete?" *Physical Review*, 47, (1935), p. 777.

27. From the book, *Albert Einstein, Philosopher Scientist*, P.A. Schilpp, Evanston, IL: Library of Living Philosophers, 1949, p. 85.

28. Reprinted by permission from J. Clauser, "Early History of Bell's Theorem," an invited talk presented at the Plenary Historical Session, Eighth Rochester Conference on Coherence and Quantum Optics, 2001, p.11.

29. Alain Aspect, "Trois tests experimentaux des inegalités de Bell par mesure de correlation de polarization de photons," A thesis for obtaining the degree of Doctor of Physical Sciences, University of Paris, Orsay, February 1, 1983, p. 1

30. The ghost image experiment was reported in Y.H. Shih, "Quantum Entanglement and Quantum Teleportation," *Annals of Physics*, 10 (2001) 1-2, 45-61.

31. The above discussion is adapted by permission from Michael Horne, "Quantum Mechanics for Everyone," *Third Stonehill College Distinguished Scholar Lecture*, May 1, 2001, p. 4.

32. "Bell's Theorem Without Inequalities," by Greenberger, Horne, Shimony, and Zeilinger, *American Journal of Physics*, 58 (12), December 1990, pp. 1131-43.

33. Reprinted from P.K. Aravind, "Borromean Entanglement of the GHZ State," *Potentiality, Entanglement, and Passion-At-A-Distance*, 53-59, 1997, Kluwer Academic Publishing, UK.

References

Much of the work on entanglement and related physical phenomena has been published in technical journals and conference proceedings. References to the important articles in these areas have been made throughout the text. The following is a list of more accessible references, which are more appropriate for the general science reader, and it includes only books that a reader may obtain in a good library, or order through a bookstore. Readers with a deeper, more technical interest in the subject may want to track down some of the articles referred to in the text, especially ones appearing in *Nature, Physics Today*, or other expository journals.

BOOKS RELATED TO ENTANGLEMENT AND QUANTUM MECHANICS

Bell, J. S. *Speakable and Unspeakable in Quantum Mechanics*. New York: Cambridge University Press, 1993. This book contains most of John Bell's papers on quantum mechanics.

Bohm, David. *Causality and Chance in Modern Physics*. Philadelphia: University of Pennsylvania Press, 1957.

Bohm, David. *Quantum Theory*. New York: Dover, 1951.

Cohen, R., S., Horne, M., and J. Stachel, eds. *Experimental Metaphysics: Quantum Mechanical Studies for Abner Shimony*. Vols. I and II. Boston: Kluwer Academice Publishing, 1999. These volumes, a *festschrift* for Abner Shimony, contain many research papers on entanglement.

Cornwell, J. F. *Group Theoey in Physics*. San Diego: Academic Press, 1997.

Dirac, P. A. M. *The Principles of Quantum Mechanics*. Fourth ed. Oxford: Clarendon Press, 1967.

French, A. P., and E. Taylor. *An Introduction to Quantum Physics*. New York: Norton, 1978.

Fölsing, A. *Albert Einstein*. New York: Penguin, 1997.

Gamow, George. *Thirty Years that Shook Physics: The Story of Quantum Theory*. New York: Dover, 1966.

Gell-Mann, Murray. *The Quark and the Jaguar*. New York: Freeman, 1994.

Greenberger, D., Reiter, L., and A. Zeilinger, eds. *Epistemological and Experimental Perspectives on Quantum Mechanics*. Boston: Kluwer Academic Publishing, 1999. This volume contains many research papers on entanglement.

Greenstein, G. and A. G. Zajonc. *The Quantum Challenge: Modern Research on the Foundations of Quantum Mechanics*. Sudbury, MA: Jones and Bartlett, 1997.

Heilbron, J. L. *The Dilemmas of an Upright Man: Max Planck and the Fortunes of German Science*. Cambridge, MA: Harvard University Press, 1996.

Hermann, Armin. *Werner Heisenberg 1901-1976*, Bonn: Inter-Nations, 1976.

Ludwig, Günther. *Wave Mechanics*. New York: Pergamon, 1968.

Macrae, Norman. *John von Neumann: The Scientific Genius Who Pioneered the Modern Computer, Game Theory, Nuclear Deterrence, and Much More*. Providence, R.I.: American Mathematical Society, 1992.

Messiah, A. *Quantum Mechanics*. Vols. I and II. New York: Dover, 1958.

Moore, Walter. *Schrödinger: Life and Thought*. New York: Cambridge University Press, 1989.

Pais, Abraham. *Niels Bohr's Time: In Physics, Philosophy, and Polity*. Oxford: Clarendon Press, 1991.

Penrose, R. *The Large, the Small and the Human Mind*. New York: Cambridge University Press, 1997. Interesting discussion of quantum and relativity issues, including commentaries by Abner Shimony, Nancy Cartwright, and Stephen Hawking.

Schilpp, P. A., ed. *Albert Einstein, Philosopher Scientist*. Evanston, IL: Library of Living Philosophers, 1949.

Spielberg, N., and B. D. Anderson. *Seven Ideas That Shook the Universe*. New York: Wiley, 1987.

Styer, Daniel F. *The Strange World of Quantum Mechanics*. New York: Cambridge University Press, 2000.

Tomonaga, Sin-Itiro. *Quantum Mechanics*. Vols. I and II. Amsterdam: North-Holland, 1966.

Van der Waerden, B. L., ed. *Sources of Quantum Mechanics*. New York: Dover, 1967.

Wheeler, J. A. and W. H. Zurek, eds. *Quantum Theory and Measurement*. Princeton, NJ: Princeton University Press, 1983. This is a superb collection of papers on quantum mechanics.

Wick, D. *The Infamous Boundary: Seven Decades of Heresy in Quantum Physics*. New York: Copernicus, 1996.

Index